Joshua Sobeto

Live Sound Fundamentals

Bill Evans

Course Technology PTR
A part of Cengage Learning

Australia • Brazil • Japan • Korea • Mexico • Singapore • Spain • United Kingdom • United States

Live Sound Fundamentals

Bill Evans

Publisher and General Manager,
Course Technology PTR: Stacy L. Hiquet

Associate Director of Marketing: Sarah Panella

Manager of Editorial Services: Heather Talbot

Marketing Manager: Mark Hughes

Executive Editor: Mark Garvey

Project Editor/Copy Editor: Cathleen D. Small

Technical Reviewer: Brian Klijanowicz

Interior Layout Tech: MPS Limited,
A Macmillan Company

Cover Designer: Mike Tanamachi

Indexer: Broccoli Information Management

Proofreader: Ruth Saavedra

Library of Congress Control Number: 2009942393

ISBN-13: 978-1-4354-5494-1
ISBN-10: 1-4354-5494-4

Course Technology, a part of Cengage Learning
20 Channel Center Street
Boston, MA 02210
USA

Cengage Learning is a leading provider of customized learning solutions with office locations around the globe, including Singapore, the United Kingdom, Australia, Mexico, Brazil, and Japan. Locate your local office at: **international.cengage.com/region**

Cengage Learning products are represented in Canada by Nelson Education, Ltd.

For your lifelong learning solutions, visit **courseptr.com**

Visit our corporate website at **cengage.com**

Printed in the United States of America
1 2 3 4 5 6 7 12 11 10

To Randy Holland, who knows nothing about audio but who taught me that—be it knowledge or goodwill—the only way to keep it is to pass it along.

Acknowledgments

There is a reason why some of us have a tendency to just say, "Thanks, everyone" instead of naming names. It's because we are like that guy at the Oscars a few years ago who forgot to thank his wife. It is telling that we all remember the incident but not so much the person involved. Anyway, if I thanked everyone who deserves it, the Acknowledgments would be a book all by themselves. So, here goes nothin'.

Friends and band mates who were along in early parts of the audio journey, including Mike Krupka, Mark Lewis, Josh Lober, Julie Prince, Mark Peotter, and Jake Kelly.

To everyone at St. Therese Parish in Alhambra, California, who trusted that I actually had a clue what I was doing on my first audio gigs.

To the pros who are willing enough to teach that they actually take my calls, including Dave Shadoan, Big Mick Hughes, Paul Owen, John Cooper, Buford Jones, Tom Young, Dirk Durham, Dave Rat, Bill Chrysler, Bob Heil, Brian Hendry, David Morgan, and Mark Dennis.

To my "team," most of whom I have had the pleasure of working with at *GIG, FOH,* and L2P: Baker Lee, Steve LaCerra, Jamie Rio, David Farinella, and the late Mark Amundson, who I still miss all the time. And to Terry Lowe, Bob Lindquist, and Paul Gallo for paying me to do it.

To Mitch Gallagher, who recommended me to write this book in the first place.

To the people who started out as "audio acquaintances" and ended up among my closest friends in the world: Larry Hall, Kevin Hill, Paul Overson, and Ken "Pooch" Van Druten.

For their invaluable help with the chapter on speaker components and enclosures, Mark Gander from Harman/JBL Professional and Chris Rose from Eminence Loudspeakers.

For their direct input in the chapters on actually doing the gig, Mike Allison from Clair/Bon Jovi, touring guy extraordinaire Mical "Mikey" Catarina, and the tribe at ProAudioSpace.com.

Finally, to my wife of 22 years, Linda (who has gone as far as buying me gear she did not understand and even joining the band when I needed a backup singer), and my daughter, Erin, who have both put up with stupid, crazy hours and a husband and dad who is often physically in the same room but somewhere else entirely in his mind, and who long ago stopped asking about the garage full of audio gear because they know the answers are, "Yes, I need it all," and, "It's gathering dust because I need to fix it."

About the Author

Bill Evans has been working in music and audio since...well, let's just say that it has been a really long time. He was the editor of *GIG* magazine in the '90s and has been the editor of *Front of House* (fohonline.com) since 2002. Bill is also the Minister of Propaganda for the Live2Play Network (L2PNet.com) and leads the tribe at the social network ProAudioSpace (proaudiospace.com).

In his "spare time," Bill has fronted the band Rev. Bill and the Soul Believers (revbill.com) since 1984. He likes Little Feat, scuba diving, William Gibson, and lobbing libertarian political firebombs on Facebook.

Contents

Chapter 10
Equalization 61

Chapter 11
Other Channel Stuff 65

Chapter 12
The Master Section 69

Chapter 13
Gain Structure 73

Chapter 14
Aux Sends and Returns 77

Chapter 15
Monitoring 83

Introduction

I still remember my first PA system. . . .

Actually, my first PA was a guitar amp. The gig was a fundraiser carnival for the Crippled Children's Society, and the year was 1973. My younger brother had succumbed to leukemia a few months earlier, and that organization had been instrumental in my family's financial survival. We took a mic from a reel-to-reel tape recorder, put it inside my nylon-string guitar, and plugged it into one side of a two-channel guitar amp; a RadioShack vocal mic went into the other. Add some drums and a trombone trying to play bass lines, and we had a band.

Fast-forward a few years, and we bought our first real PA from a guy living in one of the now-razed bungalows behind the Hollywood Bowl. He needed the money to fly to England to see his guru. Really. You can't make stuff like this up, and it was still the '70s.

It was a custom-made 16-channel mono mixer with a separate monitor output. We got the mixer in a road case, a 150-watt power amp, a pair of Altec horns, two Cerwin-Vega folded-horn bass bins, and a couple of Shure Vocal Master columns that we turned on their sides and used as monitors. We dubbed the console T.I.M.—totally intense mixer. And at that time, we had the biggest, baddest PA on our little band circuit.

Since that time I have been through pretty much every twist and turn and advance in the world of live performance audio. I was the quintessential "guy in the band who owned the PA," and for many years I set up and ran the PA in addition to playing in the band. Almost 20 years ago, I started renting out my system and mixing other bands. I learned by watching, listening, and emulating those who were farther up the audio food chain. The only training I knew about was of the on-the-job variety, and if there was a book that explained it all, I'd never heard of it.

In the late '80s—through a series of seemingly unrelated contacts and incidents—I ended up working for someone who had been in a competing band on that old circuit on a magazine for musicians called *GIG*. It was there that my education began on how to explain what "all the knobs and buttons do" to audio novices. Later—via another string of "coincidences"—I took the helm of *Front of House,* a trade magazine focused entirely on the live audio biz. Over the past almost eight years, I have had the pleasure of interviewing and learning from some of the top live audio pros in the world. I have had the honor of meeting audio pioneers including Bill Hanley, Bob Heil, Stan Miller, and a bunch of others I will regret leaving out after this gets printed.

And I have kept my hand in the more musician-oriented side of things through my involvement with the Live2Play Network and writing a series of Live Sound 101 columns for our print and online publications. My hope is that this book will serve as an extension of those articles and will give those who read it a solid grounding in the basics of live performance audio. (I remember trying to put together written instruction for my band mates on how to set up the PA for a gig I was going to be late for. When I got up to 11 pages, I started to wish they had all read a book like this.)

I have often said that I know too many really good sound guys to ever claim to be one. I'm just the guy in the band with the PA who happened to get day gigs where he got to help explain it all. But you never know where the little bit of knowledge you share will end up.

A few years ago, I was backstage at a Toby Keith show talking with his longtime front-of-house engineer, Dirk Durham. Later that evening, Dirk would be driving a huge, state-of-the-art sound system and bringing the sound to some 15,000 screaming fans. I asked how he got into the audio biz, and he told me about being a rodeo guy who had friends in bands who told him that if he was going to hang out, he needed to make himself useful. So he started hauling gear and eventually setting up the PA and finally mixing the band.

When I asked him how he learned how to set up and run the system, I expected him to tell me about some guy he met who took him under his wing and showed him all the ins and outs of audio. I was more than surprised when he said, "I learned everything I know about sound from one place. You see, there used to be this magazine called *GIG*. . . ."

If you learn anything from this book, be generous and pass it along. You never know where it will end up.

The Gear

PART

I

1 What Is Sound?

The anticipation in the room is palpable. The space is already filled with sound as the conversations of thousands of fans converge into a throbbing, living hum. Finally, the house lights go down, and the crowd explodes. On stage, all that is visible are a number of flashlights illuminating the floor. The people in the crowd crane their necks, stand on chairs, and jump up and down, hoping for a pre-show glimpse. Finally, you hear the command, "Cue sound, cue lights." You bring the main faders up as the stage lights come on, the star of the day asks the crowd how they're feeling, and the band launches into its first tune as the crowd roars. And there you are at the console, controlling it all.

Sounds pretty cool, doesn't it? And you *can* get there, but you need to learn the basics first. Starting at the very beginning: What is sound?

Good Vibrations

At its most basic, sound is vibration. An event occurs—anything from a guitar player picking a note to the proverbial tree falling in the woods. The event itself is a disturbance that excites the molecules in the surrounding air, which creates a wave that travels away from the point of the original event. That wave is going to look something like what you see in Figure 1.1. That explanation is a little on the dry side, so let's try it again. Eric Clapton peels off a blazing blues lick, which causes the molecules in the air to go into a frenzy and start bashing into each other at different speeds. The speed makes the frequency or pitch happen, and the force with which they bash into each other equals volume. Cool?

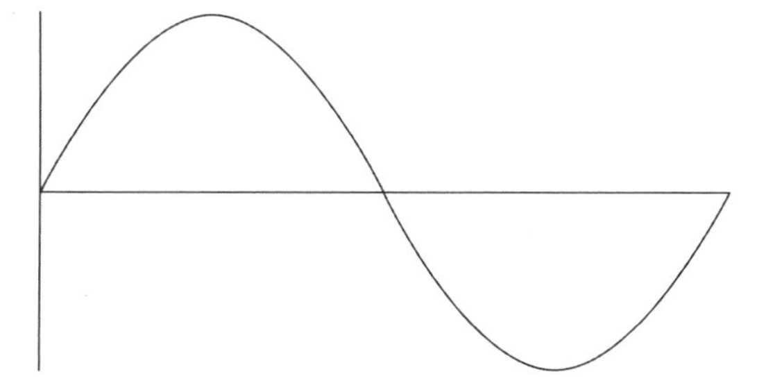

Figure 1.1 A sine wave.

The truth is that air does not have to be involved—the medium could be liquid or even solid, although there has to be some kind of medium. Sound can't travel in a vacuum.

There was an old sci-fi movie called *Alien* that used the tagline, "In space no one can hear you scream." Cheesy but true...

Depending on the nature of the original disturbance, the amount of time it takes for that wave to repeat itself can be measured in terms of distance, because the speed of sound is constant. At this point it is easiest to look at these waves as a *vibration*. The speed of the vibration—that is, the time it takes to complete a cycle as in Figure 1.1—determines what we humans call *pitch*. The faster the vibration, the higher the pitch. We measure the speed of these vibrations in cycles per second, or hertz (named for noted German physicist Heinrich Hertz). Every musical note produced by any instrument can be described in terms of frequency or hertz (see Figure 1.2).

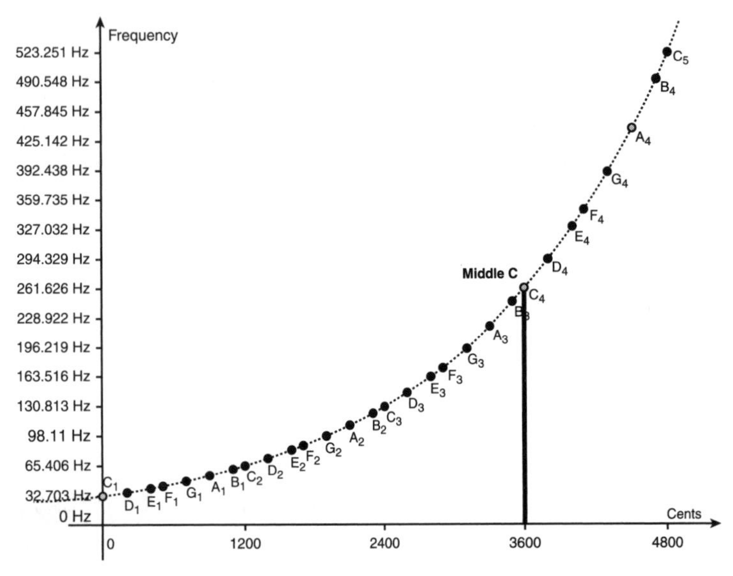

Figure 1.2 This chart shows how frequency translates to notes. Knowing this and being able to combine it with the information in the chart in Chapter 10 that shows the frequency range of various instruments will make your job of crafting a really coherent mix much easier.

An electric guitar with 24 frets can go as high as about 4800 hertz (Hz)—or 4.8 kHz—and down to about 330 Hz. A bass goes down to about 80 Hz and up to about 200 Hz. Just in case you are not into the whole metric thing, kilo means thousand. So 5,000 Hz is 5 kHz.

A couple of things to note: First, these numbers are rounded off and are not exact. Second, one of the best things you can do as a fledgling sound person is to memorize the frequency range of the instruments you work with. If you have a tenor sax whose low end is out of control, then knowing which frequencies to cut will make your job easier.

It is actually more complicated than that, because any acoustic sound source produces the primary frequency as well as a series of sympathetic vibrations at regular intervals based on the frequency of the primary tone. These are called *harmonics*.

One final note on the speed thing: Frequency has nothing to do with the speed of sound. Sound moves at the same rate regardless of frequency. Things such as temperature and humidity can affect the speed of sound, but the generally accepted number is 1,125 feet (or 343 meters) per second in a dry environment at 68 degrees F.

If you were like me, you hated math in school, and your eyes may be glazing over right about now. But this stuff is important. Understanding how sound is produced and how it moves will play into everything from designing a system for a small club or church to setting delay stacks in a stadium. Knowing frequencies will help you kill the feedback that is screaming through the wedges and threatening your employment status. So WAKE UP!

How Loud Is Loud?

Sound pressure level (SPL) is measured in terms of decibels, or dB. Why the capital B? The decibel originated in the Bell Laboratories and was originally part of efforts to quantify loss in audio levels in telephone circuits. The "bel" part is in honor of Alexander Graham Bell. (Don't ask me why they dropped one L—it is one of those mysteries of the universe.) "Deci" is the prefix that denotes 10 in measurement systems and refers to the fact that the decibel system is a base-10 logarithmic scale. If you know what that really means, then take a moment to pat yourself on the back for not sleeping through math class. I personally am not doing any self-patting, but that does not mean I can't explain what decibels mean for us sound types.

The decibel system does not really measure absolute sound levels, but rather the difference between levels. In other words, we have all agreed that 0 dB is the point at which the average human begins to perceive sound. Please note that is average *human*—not average Tool fan, whose ears are likely half blown out and whose threshold of perception is going to be substantially higher than someone whose tastes run more along the lines of Barry Manilow or Norah Jones. From that agreed-upon 0 dB, things get a little less clear. From a straight physics perspective, a 3-dB increase is a doubling of power. However, doubling the power does not mean doubling the perceived volume. Most references will set that number at 6 dB, but many audio types maintain that they do not perceive a doubling of volume at anything less than a 10-dB boost.

For our purposes, we'll go with the 6-dB standard. When it comes to volume, it is always better to be conservative. Sustained exposure to volume levels above 90 to 95 dB can result in permanent hearing loss. For someone who seeks to make a living with his or her ears, this is a crucial concept. (One A-list mixer I know was attending a concert with his wife, and after two songs he said to her, "My ears pay the mortgage. I need to leave." And they left.)

What you need to take away from all of this is the fact that 120 dB is not twice as loud as 60 dB—it is more than 1,000 times as loud. Do the math. If each increase of 6 dB is a doubling of perceived volume, then the 60-dB difference between 60 and 120 dB represents 10 doublings of the original level, or 1,024 times the original. Even using the more lenient 10-dB standard means six doublings, or 64 times as loud.

Just some points of comparison: Normal conversation (again, normal being not between two Tool fans) is about 65 dB. A gas-engine lawnmower at about 3 feet is in the 107-dB range. Your average loud concert can run in the 112- to 115-dB range, and 125 dB is the threshold of pain.

In musical terms, things don't shake out the way you think they might. I mean, a flute is quieter than a piano, right? A flute can range between 90 and about 103 dB, and a piano at normal practice volume is 60 to 70 dB. Even played *fortissimo* (very loudly), it ranges between 85 and 103 dB.

OSHA, the federal agency charged with setting safety standards in the workplace, allows for only half an hour a day of exposure to sound at 110 dB. That show you are mixing is 90 minutes long. Think about that when you are watching the meter hit 112 during the first song.

Another Kind of Doubling

Traditionally, sound was seen as traveling in an ever-expanding sphere centered on the original source of the sound. Designers of loudspeakers have made huge strides in controlling and aiming sound, as has the military. One got a weapon that can cause damage by using nothing but very focused beams of sound, and the other got the line array. And plenty of experienced sound techs will tell you that a line array *is* a weapon in the wrong hands. But that all comes later. For now, we stick with the classic physics, which gives us something called the *Inverse Square Law*. I am not going to get into the math, because all you really need to know is the result, which says that for every doubling of distance from the source, the perceived volume drops by 6 dB. There is that 6-dB figure again....

Practically, what that means is that a sound measured at 90 dB at 1 foot from the source (for our purposes, usually a loudspeaker) will measure 84 dB at a distance of 2 feet, 78 dB at 4 feet, 72 dB at 8 feet, and 66 dB at 16 feet.

So, if you are trying to get an SPL of 90 dB 40 feet away from the stage, then your level at 1 foot would have to be more than 120 dB. This is important when it comes to designing systems, both in terms of the amount of power you need to get the desired volume at the desired distance and with regard to keeping the audience a sufficient distance from the speakers to keep from hurting them (the audience, not the speakers—drunk frat boys have been known to damage speakers from distances as great as 100 feet).

From this very short introductory chapter, you should now understand how sound is created, how it travels, how loud is loud, and how loud is too loud.

2 Welcome to the Signal Chain

We have taken a basic look at the nature of sound, how it is created, and how it moves. Now it is time to get down to the nuts and bolts of setting up and running a live audio system.

The first thing to remember is that what you are doing is reinforcing the sound created by whoever is onstage. Over the years, that job has gone from just getting the vocals audible above the guitars to situations where every conceivable sound source has a mic on it. But despite the seemingly huge change, the job really remains the same: Make sure the audience can hear each voice and instrument clearly. In its journey from the source to the ears of the audience, the sound (which we will often refer to as the *signal*) passes through many stages, and we will look at each of these separately. The entire path is called the *signal chain*.

End to End

Remember that sound is energy. There are a lot of different kinds of energy, but the kind we can hear is called *acoustic energy*. The devices we use to control sound can't work with acoustic energy; they need to work with electrical energy. But, you guessed it; we can't hear electrical energy, so some transformations need to take place.

The devices that accomplish this are called *transducers*, and they convert one form of energy to another form. Both microphones and loudspeakers are transducers. The mic converts acoustic energy into an electrical signal. That signal travels through the chain getting adjusted, massaged, and sometimes plain beat up until it reaches the speaker, where that energy is converted from electrical back to acoustic.

Because these components make these crucial conversions, they are arguably the most important pieces of the puzzle. On the mic end, we operate on the garbage-in-garbage-out principle (and no, I am not referring to the talent of the person or instrument feeding the mic). We are referring to the fact that if that initial conversion of energy is poorly done, then you have very little chance of saving it. The cliché of "fixing it in the mix" is—in the case of live audio—usually a lie.

Conversely, no matter how great and pure a signal you present to the loudspeaker, if the speaker is cheap, there is no way you can get anything out of it that sounds pleasant.

In an ideal world, the mics and speakers would put out exactly what is put in. In the real world, that just does not happen. While current technology makes signal "transparency" more achievable than in the past, virtually every stage a signal passes through affects the tonal quality, timbre, or color of the sound for better or worse. One thing that separates really good engineers from their less notable counterparts is an almost encyclopedic knowledge of the gear—not only how each piece affects the sound, but more importantly, how combinations work together. That knowledge—and the ability to use their ears—is what allows them to make their act sound good on any system in any environment.

No Excuses Ken Van Druten—known as Pooch throughout the industry—has a list of clients that most engineers drool over. He specializes in harder, heavier acts and is one of the best at what he does. The first two times I heard Pooch mix were at stadium gigs with huge systems that he was familiar with, and there were no real issues with acoustics. But the third time was in a Las Vegas nightclub called Rain. The act was Kid Rock, and the room, while beautiful, was an acoustic nightmare. Round. All glass, chrome, and rock with a very high ceiling. Add to it the fact that he was mixing on a rig that was totally new to him. I am not a Kid Rock fan, but the mix absolutely rocked. I was with a sound-guy friend who is a very good mixer and knows it. He is also not shy about saying when another engineer is not cutting it. When he says, "It was one of the best mixes I have ever heard," it means something. Pooch is on the A List for a reason. He goes into any venue on any system with any artist and makes it sound as good as it possibly can. If you are going to be in this business, that's your job.

After the acoustic energy (original sound) is transduced or converted into electrical energy, it travels via either wireless transmission or a cable of some kind to a stage box, also known as a *subsnake*. This is a box with audio connectors on one end and a bundle of cables coming out of the other end. Sometimes this stage box converts the signal from analog to digital for transport over fiber or Cat-5 (computer networking) cable. Other times, it is split into two or three identical pieces, in which case this box is generally referred to as a *split*.

The console is going to be your main tool if you end as a mix engineer, but don't be fooled into thinking that this is the most important part of the system. It is important, but it lies in the middle of the chain. Remember that your most crucial components are those that do more than adjust the signal—they actually convert it.

The console consists of several parts or sections. First is the input, which is where the very small signal from the mic is goosed up to something the console can use. The piece that does the goosing is a preamplifier, also known as a *mic pre*, *head amp*, or even just *pre*.

The preamp both feeds and is part of the individual channel strips. These include tools to adjust the signal's relative volume plus its sonic characteristics. Changing the tonal quality of a signal is known as *equalization* or *EQ*. Depending on the "level" of the console, it may have anywhere from two to four or more bands of EQ.

These bands of EQ can be as simple as the bass and treble controls on your home stereo or as complex as three controls for a band that allow you to determine the center frequency, the width of the area surrounding that center frequency that is affected, and then an amount of boost or cut for the band.

The next set of controls in a channel strip is the auxiliary, or aux, sends. If you think of the channel strip as a kind of robotic assembly line, the sends are where a signal—first determined to need extra "work" not available on the console—is sent off to be worked on before rejoining the assembly line. On a high-end gig, these sends on the main console are pretty much used to send a portion of the signal to some kind of effect—reverb, delay, chorus, compressor—and then return it to the channel. If you have enough input channels on your console, you can return the effect, which allows for greater control, but otherwise you will use one of the returns, which we will get to later. Also, aux sends come in two flavors—pre-fader and post-fader.

Better consoles will also have subgroups or VCAs (*Voltage Controlled Amplifiers*), which allow you to assign groups of inputs and control them all from a single fader. For instance, you may have all of your drums, backing vocals, or a horn section on a sub or VCA. You still have the control to tweak an individual instrument in the group, but you can also bring the entire group up or down without having to muck around with seven or eight faders.

Following this is the master section, which includes the master fader as well as aux returns, recording outs, playback inputs, and your talkback mic for communication with the band onstage without screaming yourself raw.

What comes after the console depends a great deal on the size and type of system. If you are using powered, or *active*, speakers, you may very well just go from the console outputs directly to the speaker inputs with perhaps some kind of processor specific to those powered speakers in between. If you are using a more standard "passive" system, then your next stop is the drive rack.

These days, the big, heavy drive racks are disappearing fast. We used to need graphic EQs, compressors, and maybe a delay unit. Given standard sizes of gear, this could easily be a 12-space rack. Today, via the magic of digital signal processing (DSP), we can accomplish all of that and more in one or two rack spaces. Because dbx put out an actual line of products called DriveRack, we have renamed these devices *speaker processors*, and nearly everyone who makes any kind of EQ or crossover makes a speaker processor, including BSS, Carvin, dbx, Sabine, Yamaha, and probably a half-dozen that I am forgetting about right now.

A moment ago, we briefly addressed the idea of powered, or "active," speakers. Usually processes including crossover and time alignment (we'll get there—patience, Grasshopper . . .) are handled by the circuitry inside a powered speaker, but not always. For example, Meyer has designed and built their own processor specifically and only for Meyer powered speakers.

Power It Up

Once the signal has been "processed" (in other words, split into separate low-, mid-, and high-frequency signals and each of those optimized), it must be amplified. Up until this point, the signal that represents the sound is very weak and needs to be boosted heavily in order to move the speaker cone and again transduce the electrical signal back into acoustic energy. This process is done by the power amp. The signal goes into the amp and comes out much stronger but it is still an electrical signal—no sound yet.

Which is where speakers come in—the end of the chain—and that electrical signal gets goosed up to a point where it can drive a magnetic "motor" in the speaker assembly. The motor is attached to the voice coil, which is attached to a paper cone or metal diaphragm. Moving the cone or diaphragm creates an event that excites the surrounding air and—voilà!—we have sound again.

Now, armed with a basic knowledge of the signal chain and which part serves which function, we can start making some noise—and getting into each part in greater detail.

3 It All Starts with a Mic

As we agreed earlier, the format of this book follows the signal chain from beginning to end. And the first thing we need to do is to convert the acoustic energy of the original sound into electrical energy that the system can use. For that task we use a microphone. While a mic is not the entire system, it just may be the most important part.

Years ago, I was having dinner with a rep from Shure—the biggest maker of mics in the world. He told us about a call they got at customer service from a woman who said their product was defective because she took it out of the box and sang into it and nothing happened.

When it comes to mics for live sound, there is a plethora of choices, and one of the things that separates experienced sound engineers from newbies is the ability to choose the right mic for the job, be that adding thump to a kick drum or getting the choir loud enough to fill the room.

Mics come in a lot of different flavors, but for our purposes, we will limit things to dynamics and condensers. ("Yes!" he said, with a nod to his ribbon-worshipping friends, "There are some folks using ribbons"—including rockers such as Aerosmith's Joe Perry, who uses Royer ribbons on his guitar amps—and they sound great. On the downside, they are expensive and fragile—which makes them a big risk when used for the stage.) But this is not an article about mics, so here are the bare basics. A dynamic mic uses a magnet and a diaphragm. The movement of the air caused by the original sound moves the diaphragm, and the movement of the diaphragm causes changes in the magnetic field between it and the magnet. This changing field creates a varying, low-voltage signal. A condenser microphone is similar except it uses a charged plate instead of a magnet. As a result, the mic needs power to work (typically 48 volts). That power can come from a battery or an external power supply, but most often it comes from the mixing console. This is called *phantom power*, and buying a mixer without this feature is shortsighted.

When it comes to sound, general thinking says a dynamic is more roadworthy but less detailed, especially in the high end. A condenser is more fragile but puts out a more "detailed" signal. Most dynamic mics also exhibit a trait called the *proximity effect*, which causes the low frequencies to be emphasized as the sound source gets closer to the mic. This is part of the reason why sound guys bitch about "mic eaters," although being too far from the mic is just as bad (not enough energy getting to the transducer), and some artists use the proximity effect as part of their sound. Condensers also exhibit the proximity effect, but generally to a lesser degree.

The preceding is a pretty gross generalization, and a lot has changed in mic technology in the past few years. Condensers have gotten downright tough in comparison to what they used to be. In my roles as editor of both *FOH* and the Live2Play Network, I insist that every mic we review go through a drop test—at least five feet, capsule down onto a hard surface, such as concrete. We have yet to have a mic fail the test. We have dented quite a few, but they always work when we plug them in after dropping them.

Before we move forward, I see that I just used a term that we did not explain. The capsule of the mic is the part that captures the sound. It can be ball-shaped, paddle-shaped, or even capsule-shaped, depending on the intended use. The rest of the mic is called the *body*. On a handheld or vocal mic, it is a cylinder that easily fits in a singer's hand. The body of a mic made for a kick drum may look like a continuation of the capsule and give the whole thing a kind of oval shape.

A note about vocal mics: The end is usually ball-shaped, although that ball may have a flat end. Under the usually steel mesh screen is some foam. It can be an integral part of the screen or a separate piece, depending on the mic. This is called a *wind screen* and serves mostly to dampen plosives—heavy sounds emanating from the singer that move a lot of air and could damage the diaphragm. If you are really looking to dampen the effects of actual wind, there are big hollow foam balls with a hole in one end that fit over the top of the mic for that purpose.

What's the Address?

I once did a gig for a church play that included an actor paying the part of a radio announcer. We used a condenser mic mostly because it looked right. We would set it up, and every time he sat down to use it, he would reposition it so that he was speaking into the top of the mic, because everyone knows that's how you use a mic—right?

Well, not really. It depends on the position of the diaphragm. The two arrangements are generally referred to as *top-* or *front-address* and *side-address*. The mic in our example was a side address, so the actor was making it very unlikely that anyone would hear him because the diaphragm was facing the desk and not his mouth. (It also caused us feedback problems.) The majority of the mics you will use in live sound will be top/front-address but not always. And you need to know before you start setting them up.

A couple of years ago, I was doing a gig attached to a pro audio tradeshow sponsored by a speaker manufacturer. It was a weird gig at a club across from the convention center with two bands—my 10-piece soul review and a very good Ozzy Osbourne tribute. Oh, and it was a really loud gig. (Remember, it was put on by a speaker company to show off their system, which generally means cranking it up pretty hard.)

At sound check we were having feedback problems, and the assumption was that it was a monitor issue. (My band is usually in-ear, but this gig was all wedges.) But the sound company owner—who was trying to stay out of it and let his staff take care of things—knew I was using a condenser mic known for being very hot and pretty wide in its coverage pattern. The main P.A. was a line array also known for a wide pattern, and after about 20 minutes of trying to find the feedback in the system, he walked onstage, unplugged my mic, and replaced it with a very narrow dynamic mic—and the squealing magically disappeared.

It's All about Heart

When you are in the studio, you will find mics with many different pickup patterns, including figure-8 and omni. (The former picks up sound equally from the front and rear and rejects sound from the sides, while the latter picks up equally from every direction.) But onstage—except for specialized applications such as choirs and some orchestral uses—you will find almost 100 percent of mics to be of the unidirectional type. You would think that unidirectional (meaning one or a single direction) mics would pick up sound from only one direction. But it is not that simple. What you actually get are several flavors of cardioid, or heart-shaped, pickup. The basic cardioid pattern looks something like you see in Figure 3.1.

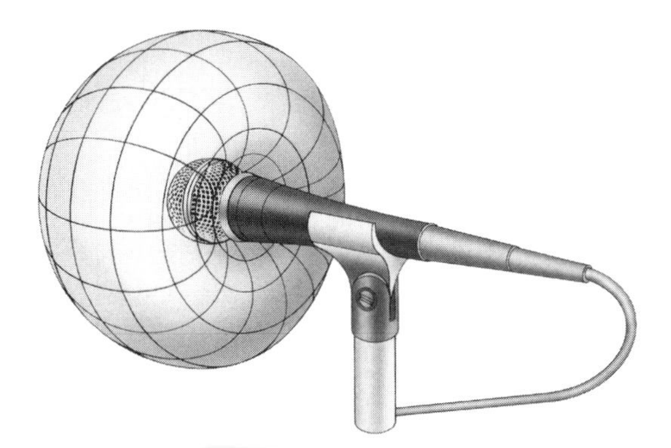

Figure 3.1 A cardioid pickup pattern. Image courtesy of Shure Inc.

As you can see, at zero degrees (or straight on), you get the full response of the mic, and it gradually falls off and dips to a theoretical level of zero at 180 degrees. The idea is to get the sound you want into the mic and reject the stuff around it. But it only works so well. Look at that plot again and notice that at 60 degrees off axis, the mic is still picking up 75 percent of what it does from the front. And for a very long time, this was the norm.

But new technology—especially new materials for the magnetic structure of the mic—allowed for a tighter pattern known as *super-cardioid.* Figure 3.2 shows what it looks like.

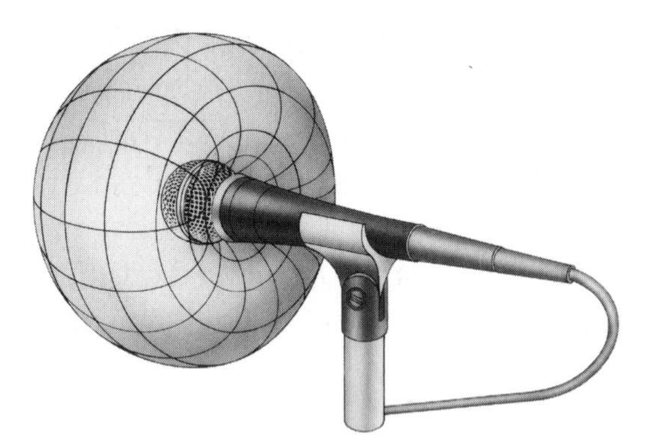

Figure 3.2 Super-cardioid pickup pattern. Image courtesy of Shure Inc.

As you can see, the response of this mic falls off a lot faster as you move off center. But nothing comes free. Look at the bottom of the plot, and you will see that at 180 degrees, the response is actually much stronger than the standard cardioid. So if you are using standard wedges for monitors, they need to be placed at an angle and not facing the performer straight on. An even tighter pattern known as hyper-cardioid is also available. It is tighter off-axis but has an even larger lobe at the back of the mic. With a super- or hyper-cardioid, straight monitors mean more feedback—exactly the opposite of a standard cardioid.

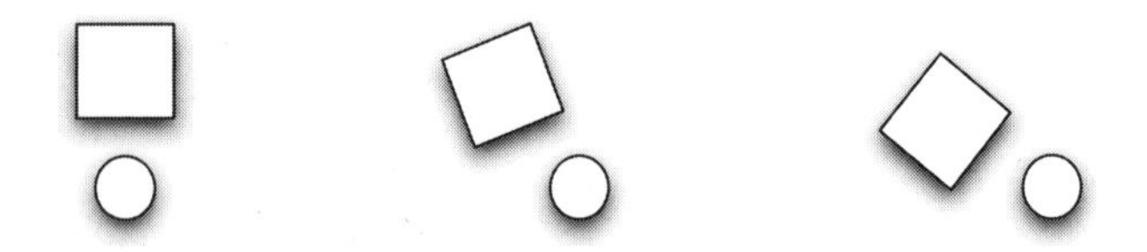

Figure 3.3 Three examples of monitor placement. The circles represent the mics, and the squares represent the monitors. In the first, a standard cardioid mic is being used, and the monitor is placed directly to the rear of the mic. In the middle example, a super-cardioid mic is being used, and the monitor is placed just off center in order to avoid the lobe at the rear of the mic's coverage pattern. In the third example, a hyper-cardioid mic is being used, and the monitor is placed even more off center to avoid the hyper's larger rear lobe. Remember, it is all about putting the monitor where the mic is least likely to "hear" it.

Although they are rarely used in a live setting (with notable exceptions of Danny Leake with Stevie Wonder's percussionists, for example), Figure 3.4 shows an example of an omni-directional mic. Notice that it picks up from any point around the mic.

So What Does It All Mean?

A couple of things. First, vocal mics come in two basic flavors—dynamic and condenser. What you need to know on a practical level is that condensers are generally thought to sound more open and airy than dynamics, and they generally provide a more

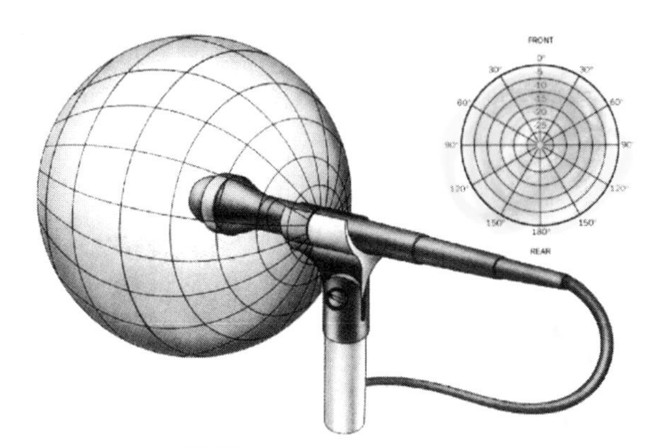

Figure 3.4 Omni-directional pickup pattern. Image courtesy of Shure Inc.

detailed sound. But for a long time, they were not suitable for live use for two reasons. First, they were fragile. Drop one, and it would likely not work afterward. Second, they have a wide response area—at least as wide as a typical cardioid dynamic.

But two things have changed that have made condenser mics pretty common, especially among lead vocalists. First, they have become a lot more roadworthy, and second, the move toward in-ear or "personal" monitoring has greatly lessened the possibility of feedback from a mic with a wide pattern. (Note that this gets more complicated, as some very smart people are doing actual new development in the mic field. Specifically, live-sound legend Bob Heil has released a line of dynamics that sound—by all reports—at least as good as most condensers and have a much tighter pattern.)

So what is the bottom line? It really depends on stage volume. On a loud stage, you need a tight pattern, and that generally means a dynamic. (One sound guy I know who mixes a very big Nashville act calls one of the standard industry condenser vocal mics the *moving drum mic* because it picks up so much drum sound in the vocal channel.) If you have a quiet stage or personal monitoring, you may be able to enjoy the generally higher-quality sound and greater detail of a condenser. What do I use? My own band is on personal monitors, so lead vocalists get condensers (a mix of Shure, Audix, Audio-Technica, and AKG, depending on the gig), but my mic locker contains plenty of very tight dynamics as well, for those gigs where a condenser is too wide. As always, when deciding what to buy, it comes down to the eternal question: "What are you going to use it for?"

A Place for Everything and Everything in Its Place

Where you place the mic in relation to the source of the sound has a huge impact on the final result. Distance from the source and angle (on-axis or off-axis) are the two things you are most concerned with. We will go over some very basic principles, but the truth is that this is a place where experimentation and experience rule the day. The "old dude" on the crew may not know the minutiae of every digital console out there, but he sure knows how to mic a kick drum so that it feels like it is kicking you in the chest without sounding muddy. Watch. Listen. Learn.

The Snare

Before we get into actually miking drums, make sure this truth is drilled into your head. *Nothing* will make more of a difference in your drum sound than a good, well-tuned drum kit. If you are working with a crappy kit that is poorly tuned with worn heads, your chances of getting a good sound are about nil. At that point it becomes a case of minimizing the possible damage to the overall sound.

The most popular type of mic for the snare is a cardioid dynamic with a presence peak, but many engineers prefer the transient response of a condenser. For years, the go-to drum mic for snares and toms was a Shure SM57, but pretty much every mic manufacturer makes a mic appropriate for snare. Again, experiment.

As far as placement goes, there are many good systems out there that allow you to attach the mic directly to the drum without using a stand, which makes for a much less cluttered look onstage. Start by placing the mic about one inch in and one or two inches above the head, with it angled toward the spot where the drummer tends to hit and far enough away from the hi-hat to avoid picking up the rush of air that happens when the hat closes. Different drummers will require different placement. Although it used to be just a studio thing, it is not uncommon to see an act with plenty of money and lots of input channels on the console mike both the top and bottom heads of the snare drum, with the microphones in opposite polarity. A mic under the snare drum gets the metallic edge of the actual snares, which pairs nicely with the fuller sound of the mic on top of the drum.

The Hi-Hat

In situations where channels or the number of mics available are limited, it is not uncommon to forego a hi-hat mic and depend on the snare mic and cymbal overheads to take care of it. Situations where you do use a separate hat mic call for a condenser placed about six inches above and pointing down. (Make sure to place it where the drummer is not going to actually hit the mic.) Some engineers prefer to place them from the bottom and flip the polarity. (We'll get into polarity when we talk about channel strips.)

Tom-Toms

When miking the toms individually, the type and placement are very similar to the snare, with the mic perhaps a bit closer to cut down on leakage. (We will get into gating drum mics when we get to the processing part of the signal chain.) Again, if you are working a club or another situation with a limited channel count, you can "cheat" by placing a single mic between pairs of toms. This means having the mic farther from the drums so that both are picked up, which can negate the proximity effect that many engineers use to achieve a fuller sound. Mini condenser mics are also becoming more popular on toms. Besides the sound itself, you need to consider leakage from cymbals when placing tom mics, aiming the deadest part of the pattern toward the cymbals.

Kick Drum

Let's get one thing out of the way right off the bat—the kick drum is *not* the lead vocal. Yes, it is an important part of the foundation of a good overall sound. But I have seen far too many mixers (both experienced and not) spend more than half of their sound-check time fiddling with placement and EQ on the kick drum. Remember, 90 percent of the audience is there to hear the singer sing the songs. If the vocal sounds great, you are halfway home.

A popular mic for kick drum is a large-diameter, cardioid dynamic type with an extended low-frequency response. But wait, here is another case where lots of channels and lots of available mics open up options, and two mics on the kick is a popular option. The idea is to pair a dynamic with a condenser, with the dynamic picking up the thump of the beater and the condenser picking up the tone of the shell. Great idea, but making it work can be complicated. Most smart engineers who use two mics have rigged some bar and clamp system where the drum mics live inside the kick drum so they don't have to worry about finding the proper placement. Anytime you use two mics on a single source, you take the chance of the mics being out of phase and certain frequencies canceling each other out. Remember the picture of the sine wave with its peaks and valleys? Imagine two waves where one was at the peak at the same time the other was at the valley, as you see in Figure 3.5.

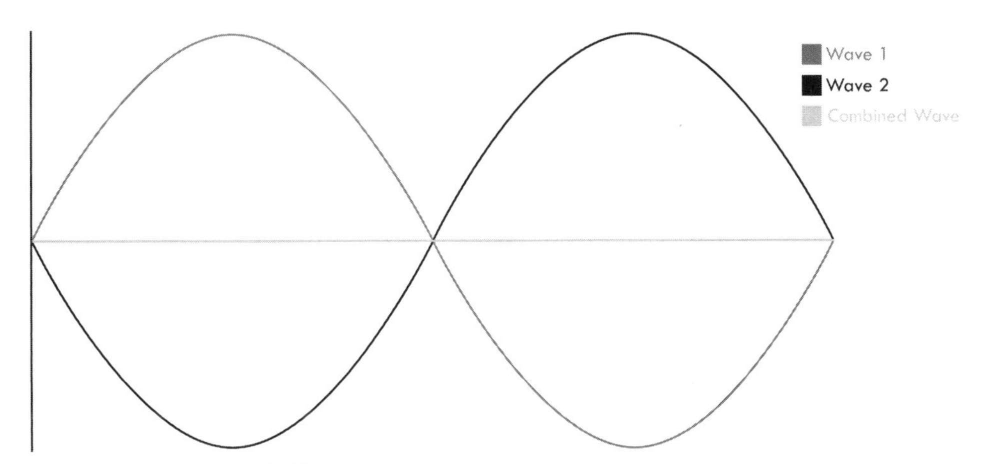

Figure 3.5 Two sine waves on the same frequency but out of phase. See how one wave is at the top of its path while the other is at the bottom? When this happens, the two waves cancel each other out. Illustration by Erin Evans.

Very small changes in placement can affect that phase relationship and drastically affect the sound, hence the "get it right and leave it there" approach. Another way to get the two-mic result without the time and hassle of getting the relative placement together is to use the Audio-Technica AE2500. This gem actually contains both a dynamic and a condenser element in the same housing with separate outputs for each, giving you all of the sonic advantage of two mics with none of the phase issues.

Cymbals

A pair of condenser mics with fairly open patterns on booms above the kit is the most common setup. But there are other approaches. Jeff Rasmussen, who at the time of this writing had been with Michael McDonald for well over a decade, has been known to use a pair of large-diaphragm condensers mounted in front of the kit. And, again, when channel count and mic availability are no object, you can get really involved. Big Mick Hughes, who has been mixing Metallica since they were a "baby band," uses what he calls *underheads* with a separate condenser mic for each cymbal. The combination of lower gain on each mic and proper gating makes for a much cleaner drum sound on what can be a very loud stage.

In addition to miking cymbals with the "underhead" approach, you can also use X-Y, overhead, and spaced-pair miking techniques. The X-Y technique uses two matched microphones. Certain microphones, such as small- and large-diaphragm condensers, are available in matched pairs. Matched pairs have consecutive serial numbers mainly so there are minimal, if any, sonic differences between them. Some even come with charts to show their responses. These matched pairs of microphones are placed next to each other with the diaphragms facing 90 degrees from one another, with the center between the two diaphragms facing the source. With the mics facing this way, one mic is panned left and one right, creating the stereo image. There are also a couple "stereo" mics that have this technique built in, such as the Shure VP88.

Next is the overhead, also known as the *spaced-pair,* technique. This is very similar to the underhead technique, but the mics are placed over top of the cymbals instead of underneath.

One thing to consider when putting mics in place, especially for cymbals, is the 3-to-1 rule. The 3-to-1 rule is as follows: When using multiple microphones, the distance between microphones should be at least three times the distance from each microphone to its intended source. If the mics are placed too closely to one another, phasing (the unpleasant kind) will occur, and when listening to both mics together, it will almost sound like the sound is in a small tunnel. An easy way to try this out is to take two mics of the same kind—say, an SM58—and turn them both up equally and put one in each hand. Talk into one mic and then start to bring the second mic closer to the first one.

On a side note, the 3-to-1 rule does not apply to X-Y miking techniques. The X-Y technique will create phasing, but that is part of what makes the unique "stereo" image as well.

Keep It Simple

On quieter gigs—especially acoustic and jazz groups—you can get away with a very minimal approach. I have seen major jazz artists mixed with just a pair of overheads or perhaps two overheads and a mic on the kick. On a rock gig, I once used a kick mic and a PZM (or pressure zone) or boundary mic (typically used as floor mics in stage productions and not in concert settings) mounted to a two-foot-square piece of Plexiglass and hung above and behind the drummer on a boom stand. It worked great, and I only used two inputs.

Guitar and Bass

The rule of thumb is to put the mic where it sounds best. With more open space around a speaker cabinet than in a drum kit, your options are more open. A couple of tips...

Although an off-axis placement can cut the proximity effect and allow the guitar to sit in the mix better, few things make me crazier than the guitar player who shows up to the gig with his own mic and simply hangs it over the top of the amp so the body of the mic is parallel to the speaker. This means that the amount of energy hitting the diaphragm dead on is somewhere between zero and none, as it all moves across the diaphragm instead of into it. Often you will find this in club situations where the band is providing their own gear, and it is generally just laziness. (It means not having to carry an extra boom stand for the amp.) Always keep a couple of spares handy for this kind of situation.

With more and more guitarists using modeling amps (which are basically big, heavy computers), there have been a greater number of players running direct. (A direct output is an XLR connection on the amp itself that goes directly to the PA without a mic. These are a fairly new development with guitar amps. Bass amps have had them as a standard feature forever.) Although a direct input theoretically captures all of the tone being produced, they can sound thin. The speaker itself is a major component of the overall sound. If you have a situation where the guitar is routed directly to the PA mixer along with the amp being miked, the main concern is phase. If the sound seems hollow, traditionally you have had two choices. One, move the microphone around until the sound is more solid. Or two, engage the phase reverse or polarity button on your mixer. But recently, Radial has come out with a device that takes both the mic and the direct (or two mics in the case of something like a snare drum) and, instead of the all-or-nothing approach of a phase reverse on one input, you can "dial in" the phase until you find that sweet spot. This is a great tool.

Horns

There are myriad options here, and rest assured that almost every mic out there has been tried. A few guidelines: A horn mic has to be able to take high SPL without distorting. The level of sound coming out of a trombone or sax can be as much as a kick drum. Personally, I am a fan of the clip-on horn mics that nearly every manufacturer makes that are purpose-built for miking horns. They solve two problems. First, you can be relatively sure you are using an appropriate tool, and second, you don't have to worry about the player's mic technique. If the player moves, the mics moves, too.

Keyboards

Most keyboards will run direct into the PA via either a submixer or a direct box. The major exception is the acoustic piano, and miking an acoustic piano is an art unto itself. For tips, read the trade magazines and look for interviews with mixers such as Tom

Young (Tony Bennett) and David Morgan (Bette Midler), who have to mic acoustic pianos for every show. There are also systems including those by Helpinstill and Earthworks that are specifically designed for pianos.

The other big exception is the "real" organ (not a synth) with a Leslie rotating speaker cabinet. These are usually miked using three inputs—one on either side of the rotating horn and another lower one in front of the rotating drum in front of the low-frequency driver.

But back to the direct box thing... No matter what level you are working at, you should always have several direct boxes in your workbox. These small boxes can be passive or powered, and they convert the instrument-level signal of the keyboard (or acoustic guitar or violin with a pickup, for that matter) that comes out of the instrument on an unbalanced 1/4-inch connection into a line- or mic-level signal on an XLR that can be jacked into the PA. It is all about cable length. An instrument puts out a high-impedance signal, which can travel maybe 20 to 30 feet before you start losing signal. A mic signal is low impedance and can go more than 300 feet without noticeable signal loss. With the increasing use of computers and iPods playing tracks to augment the actual sound coming off the stage, there are now specialized direct boxes like you see in Figure 3.6.

Figure 3.6 The Rapco LTI 100 is one of several purpose-built direct interfaces. This one is for getting a computer, iPod, or other device with an 1/8-inch stereo output into the PA on two balanced XLR connections. Image courtesy of RapcoHorizon.

Now we have all of our sound sources set up, and it is time to get them into the PA, which leads us to the next link in the chain—cables, snakes, splits, and wireless.

4 Cables and Connectors

So, you have your original sound or acoustic energy, which you have converted to electrical energy via a transducer (also known as a mic). The next step is to transport that signal to a place where it can be manipulated.

I am sure this sounds so simple that some of you are wondering why we are devoting a whole chapter to it. The truth is that not only are there myriad cable and connector types, and you need to know which one is right for which job, but also, when a system fails, it is almost always a cable or connector that is the culprit.

I used to work with a guy who played trumpet in my band but who was also a killer guitarist. He used to say that whenever there was a guitar problem, it was always the cable or the B string. Truth is, he was right more often than not, but his joke points out a stark fact—cables count. Cables are pretty much defined by construction and connectors. The combination of those things dictates what the cable is used for. Let's start with the cable part of it.

Shielded versus Unshielded

Let's start by taking a look at the simplest cable—your basic unshielded speaker cable (see Figure 4.1).

Figure 4.1 Standard speaker cable. Image courtesy of RapcoHorizon.

As you can see, what we have here are two insulated wires, or *conductors*, inside of an insulated sheath. The sheath is usually rubber or a rubberized plastic, but some high-end cable makers have started to use fabric. Rest assured that if the covering is some kind of fabric, the cable costs twice as much as anything standard.

The reason these are known as unshielded is that they consist of two conductors in a sheath. Simple. A shielded cable also has two conductors inside a sheath, but the ground conductor is not insulated and is wrapped or woven around the "hot" conductor. Sometimes the shield is made of foil. The reason for the shielding is that a long cable is basically a big antenna. When the signal on the hot conductor is small (instrument, mic, or line level), outside interference from a number of sources, including radio waves or any kind of magnetic field, can overwhelm the hot signal, resulting in noise, dropouts, and even having the local radio station being broadcast through a guitar amp or PA.

The reason unshielded cable is often known as a "speaker" cable is because the strength of the signal traveling between the power amp and the speaker cabinet is such that those kinds of extraneous signals are unlikely to cause any interference. You may hear that you should never use shielded cable to carry signal between an amp and a speaker. The truth is that in a pinch you can get away with it, but assuming you are using a balanced cable with two conductors inside the shielding, it is possible (though unlikely) that the two conductors could make contact with the shield, shorting the connection and blowing your amp.

In the case of a cable with a single conductor, the issue is one of cable size or gauge. The inner conductor on a guitar-type shielded cable is very thin—probably 22 to 24 gauge. It carries a signal as strong as what goes between an amp and a speaker—especially over long distances. When a light cable is used to carry a large signal, it is going to heat up. It is unable to carry that much signal, and the energy it cannot carry is converted to heat. (Remember our talk about transducers and energy conversion? Energy cannot be created nor destroyed. However, it can be converted from one form to another. In this case, the excess electrical energy is converted and dissipates as heat.) The bottom line is that some of the energy being produced by your amp is being wasted, and if the cable heats up enough, the insulation could melt, and then we're back to the shorting-out-the-amp thing.

The two most common connectors are the 1/4-inch (which comes in two flavors) and the XLR (which comes in two "genders").

A quick side note: I am really trying to keep this simple, but it is more complex than it appears on the surface. Just remember that this is an area where you really get what you pay for. When it comes to premade cables (yes, a lot of us "roll our own"), names such as Whirlwind, Monster, Planet Waves, Link, Rapco, or Horizon are always safe. Just stay away from molded-on connectors and look for some kind of strain relief at a minimum. If you are not sure what these terms mean, ask before you buy.

A 1/4-inch connector is what many people refer to as a "guitar" cable, and it looks like what you see in Figure 4.2.

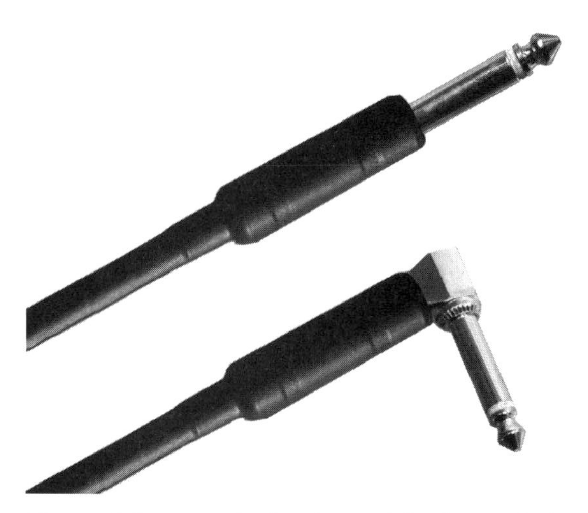

Figure 4.2 A typical guitar cable shown in both straight and 90° angled versions. Image courtesy of RapcoHorizon.

Figure 4.3 shows another variation.

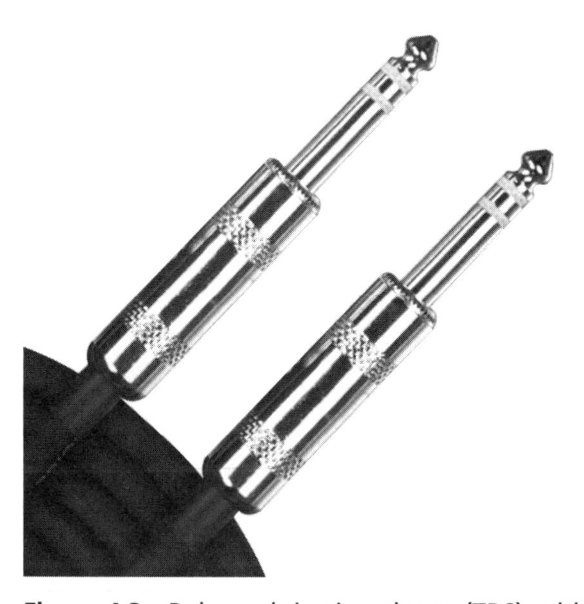

Figure 4.3 Balanced tip-ring-sleeve (TRS) cable. Image courtesy of RapcoHorizon.

While the two look very similar, there is an important difference. Look at the shaft of the first example, and you will note a single line separating it into two parts, whereas in the second there are two separators and three parts. These separating lines are insulators, and each part of the shaft corresponds to a different conductor in the cable. Sometimes these are called *mono* and *stereo*, but more accurate is TS (tip-sleeve) and TRS (tip-ring-sleeve), or unbalanced and balanced. Mic XLR connectors are three-pin and balanced, with the difference being male (see Figure 4.4) and female (see Figure 4.5).

Figure 4.4 Male three-pin XLR connector. Image courtesy of Neutrik.

Figure 4.5 Female three-pin XLR connector. Image courtesy of Neutrik.

A two-conductor or tip-sleeve or unbalanced connection is easy. One wire carries the signal, and the other is ground. (If you don't know what those two terms mean, it's time to bone up on basic electrical knowledge, which is not what we are doing here.) The "hot" or signal wire attaches to the tip and the ground to the sleeve. Simple. So why three conductors and a balanced connection? Let's ask Wikipedia.

> Balanced audio connections allow for the use of very long cables with reduced introduction of outside noise. A balanced audio connection has three wires. Two of these are used for the signal, of opposite polarity to the other. The third wire is a ground and is used to shield the other two. The signal is the difference between the two signal lines. Much of the noise induced in the cable is induced equally in both signal lines, so this noise can be easily rejected by using a differential amplifier or a balun at the input.

> The separate shield of a balanced audio connection also yields a noise rejection advantage over a typical two-conductor arrangement such as used on domestic hi-fi, where the shield is actually one of the two signal wires and is not really a shield at all, but relies on its low, but in practice not zero, impedance to signal ground. Any noise currents induced into a balanced audio shield will not therefore be directly modulated onto the signal, whereas in a two-conductor system they will be.

Okay, in non-geek this means that the noise in an unbalanced cable is canceled out in a balanced cable. The truth is that I would use a balanced cable anytime I had a choice. Every cable in a rack, every input into a mixer, and every line from a mixer to a processor or amp should be balanced whenever possible. And choose an XLR over a 1/4-inch whenever you can, just because they tend to be more durable.

But Wait, There's More

Wouldn't it be nice if there were only two connector types to deal with? But no such luck. While 1/4-inch TRS and unbalanced and XLR are the ones you will use most often, there are others that are very important.

First up is the RCA or phono connector (see Figure 4.6). Yes, this is the kind of connector you have seen on the back of your home stereo. On a real pro system, you will rarely see an RCA connection, but for club and smaller band gigs, churches, and even small theatres, you will often see a smaller mixer with RCA connectors that say Tape In and Tape Out that are used to connect consumer-grade CD players, tape decks, and so on to the board. (There are a few out there already, and by the time you read this, expect to see more consoles with iPod dock connections right on the face of the mixing surface.)

Figure 4.6 An RCA connector, also known as a *phono* connector. Image courtesy of Neutrik.

Oh wait, did I say the lowly RCA was not a "pro" connection? Silly me. That was the case for a long time, but certain digital gear—including smaller and early digital consoles—may include RCA connectors for something called S/PDIF (Sony/Philips Digital

Interface format). This is a digital format that sends two channels on one cable and receives two on another. You will also see XLR connectors meant to carry digital signals in what is called the AES/EBU format. The big difference between these and analog connections using these connectors is the quality of cable needed. Don't even think about using a standard AV RCA cable or a standard mic cable; you need to use cable that is approved for digital signals, which usually means a lower-impedance, higher-quality cable.

What do you use them for? Most S/PDIF and AES/EBU connectors are used to transfer digital signals from something such as a DVD player or computer to the console, keeping it in the digital domain as long as possible. In the case of AES/EBU, it is also found at the output of consoles and inputs of speaker processors and amps. Again, the idea is to keep the digital signal digital for as long as possible. Another digital connection you will sometimes see in pro gear is the ADAT or Lightpipe connector. If you are at Best Buy and looking for such a cable for your home theatre system (which also uses this connection), it will almost surely be called a *Toslink*. But there is not an audio pro alive who does not know about the Alesis ADAT digital recorder, and it was the first piece of gear that I know about that used this light-over-glass-fiber optical connection to send up to eight channels of audio on one cable. So don't call it a Toslink on the gig and expect anyone to know what you mean.

Weird Stuff and Power

There really was a time not so long ago when all we worried about was the standard Edison electrical plug, 1/4-inch and XLR. But those days are long gone, and there is more you need to know about. First, let's get power out of the way. Most of your gear—especially in clubs and smaller gigs—will use standard Edison AC. But the cable is rarely attached permanently to the box. It is usually removable and is almost always what we call an *IEC cable*. Figure 4.7 shows what an IEC cable looks like.

Figure 4.7 A standard IEC electrical connector.

One end attaches to the gear, and the other attaches to the electricity. There are two other kinds of power connections. First, there's the hated wall wart, which looks something like what you see in Figure 4.8, or there's the "brick on a rope," which you see in Figure 4.9.

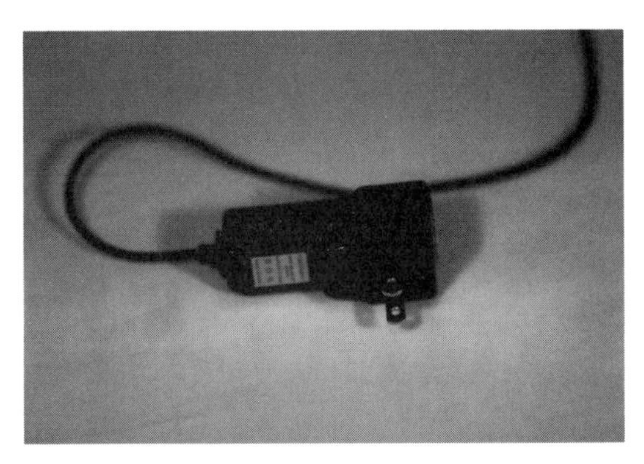

Figure 4.8 Wall warts are AC adapters that plug directly into an outlet. We hate them because they are heavy and have a tendency to fall out of the outlet in transport, and they often block one or more additional outlets on a power distribution unit.

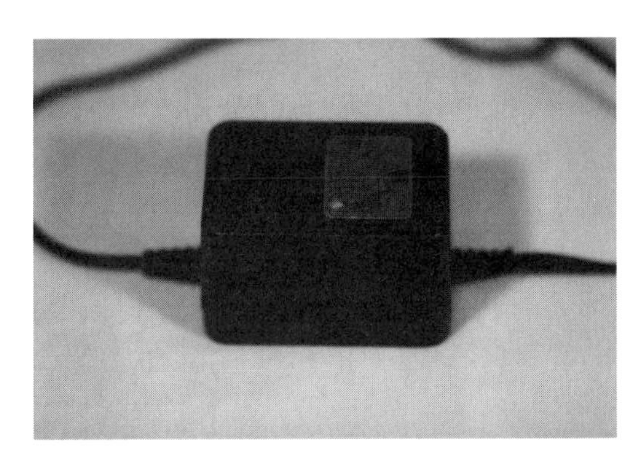

Figure 4.9 Another kind of AC adapter, known by many as a "brick on a rope." Note there are two cables protruding from the box. One carries power from the Edison outlet, and the other the converted power to the unit it is intended to work with. This is better than the wall wart, but it is still widely hated.

Either way, they plug into the 110v electrical service and convert it to something the gear needs. Usually, but not always, this is a DC—direct current—signal.

If you don't know the difference between alternating and direct current, you need to do some boning up. But it is worth doing if only for the tales of the war between Thomas Edison (alternating) and Nikola Tesla (direct) over the future of electricity use.

Why use these annoying wall warts instead of internal power supplies? Like everything else, it is all about the Benjamins. When you see that familiar UL-approved logo, it means that a company called Underwriters Laboratories has approved the gear, which means it can be insured. But getting UL approval is very expensive ($10K+ per approval), and the *only* thing they look at is the power supply. So if you are making inexpensive gear and most of it uses the same power supply, but you're making it external to the box, you do *one* UL approval instead of one for each unit that uses that power supply. So look at them as a way to save you money.

On real high-end pro gear, you may find a different kind of power connector generically called a *twist-lock* but more properly known as a *powerCON*. (Neutrik, a connector manufacturer in Lichtenstein, has a patent on the powerCON, so others you see are probably knockoffs.) A powerCON looks like what you see in Figure 4.10, and the advantage is that the connector inserts, twists, and locks into place so it cannot be accidently removed.

Figure 4.10 A powerCON receptacle. Note the slot. The connector can only be inserted one way, and once inserted, a twist locks it in place. Image courtesy of Neutrik.

On things where you often see a bunch of powerCONs used at once, such as power amps or powered-line array speakers, you will often see daisy-chainable powerCONs. The power enters one box, and then all of the other boxes are connected to the others via powerCON cables, and the power moves up the line with a single connection to the source. It makes for much cleaner cabling.

Finally, we have data connectors. Yes, I said data. Audio is becoming more and more like computer networking. In fact, a couple of months before writing this, I interviewed Tony Luna, who was doing monitors for Aerosmith, and he said, "We have gone from being audio guys to being network managers." And there is more than a little truth to that.

The first data connection we all had to deal with was the MIDI cable, which is still in use. MIDI was a protocol developed by a group of synthesizer makers who wanted their instruments to be able to talk to each other. And if you really look at the history, MIDI was the beginning of the change that turned us all into computer geeks instead of straight audio geeks. MIDI stands for *Musical Instrument Digital Interface*, and a connector looks like what you see in Figure 4.11 or 4.12.

Figure 4.11 Five-pin MIDI connector. Image courtesy of Neutrik.

Figure 4.12 Seven-pin MIDI connector. Image courtesy of Neutrik.

The only difference between the five- and seven-pin versions is that the seven-pin can carry power as well as data. Again, cleaner cabling. MIDI started as a way to use one keyboard to play the sounds in another. This led to the sound module, which was basically the synth without a keyboard that connected to another synth or keyboard controller to provide more sounds with less stuff. But soon, things such as effects units and even mixers had MIDI connections that were used to change "patches" on a piece of MIDI-enabled gear—for example, changing the type or level of reverb mid-song. MIDI continuous controllers allowed things such as fader moves to be sent to MIDI gear—you could increase or decrease volume without ever touching the actual audio or speed up or slow down a delay setting remotely. There are some small mixers meant to do double-duty, such as the Allen & Heath ZED, which is a real live-sound mixer, but each fader is also a MIDI CC device, and the board can be used to mix recordings on your computer as well—again, with everything remaining digital and nothing ever touching the actual audio.

The next data connection is one you will be familiar with if you have ever plugged your computer into a network. The connector itself is called an *RJ-45*, but they are usually referred to by the name of the cable type, which is Cat-5. These are becoming more and more common as audio is transported digitally between the console and the amp or speakers (or from the stage to the console). There are some units that on this kind of transport use specialized fiber connections, but they are not standard. Mostly what you will see is the good old Cat-5. See Figure 4.13.

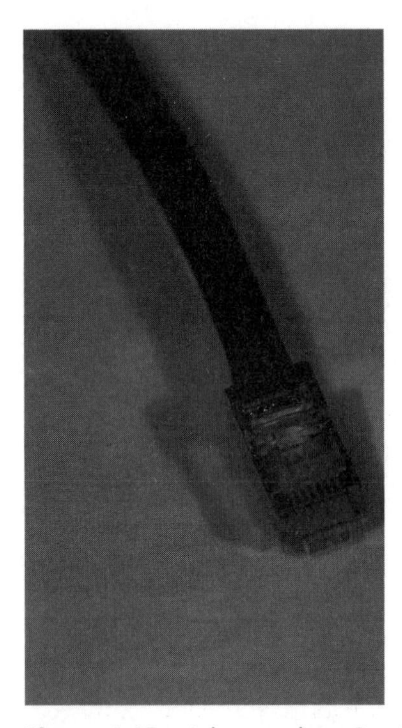

Figure 4.13 A heavy-duty Cat-5 cable.

But Cat-5 is an issue. It was designed to be used in computer networks, so it is fine in a permanent installation, such as in a church, club, or theatre, where it is plugged in once and then only unplugged if there is a reconfiguration of the system or for trouble-shooting. But for touring or the kind of local sound gigs you are most likely to do? The RJ-45 was only designed to be plugged and unplugged a limited number of times, and that is not a big number. So, they break all the time. You have a few options. The first is to carry lots of extra Cat-5, because if you don't have an extra, one will fail. (If you only have one extra, two will fail. . . .) What usually happens is that the tab breaks off.

If you don't want to carry a bunch of extra cable (and you should be carrying at least a couple extras anyway), then buy a *good* crimper and a bag of connectors at RadioShack and learn how to snip off the end and crimp on a new connector. (This is not as easy as it sounds. I have a crimper and have done this, but I'm no good at it and fail more often than I succeed.)

Another option is to only buy gear that uses the Neutrik etherCON connector, which looks like what you see in Figure 4.14. Now that RJ-45 is protected by a steel shell and looks kind of like an XLR for the outside. These rock and will last a long time. The problem is that they are significantly more expensive and take more room than a "standard" RJ-45, so not enough manufacturers are using the female end in their gear.

Figure 4.14 etherCON protects the RJ-45 connector. Image courtesy of Neutrik.

Recently, TMB made a very nice shell protector that they say will protect a standard RJ-45 for a long, long time (see Figure 4.15). It protects the tab and even has an optional cap, so as your less educated crewmates are dragging the cables across the ground, you don't even have to worry about dirt getting in the connections.

I use a poor-man's version of this but will likely switch to the TMB just because it is so much more pro. But I use cheap RJ-45 "couplers" to attach the cables to each other for moving and storage.

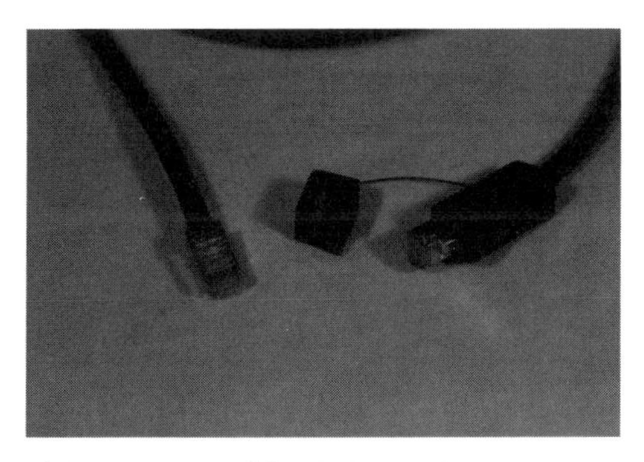

Figure 4.15 This TMB product protects the RJ-45 but does not require a different connector on the other end.

Okay, one last "data" connector, which will lead us to our next subject. (Don't you love how that works? It's called a *transition*!) The BNC connector is found most often in digital gear for word clock connections. (Again, this is *not* a tome on digital audio. If you need to know about word clock, bit rates, jitter, and dithering, try *Sound Advice on Digital Audio* [Course Technology PTR, 2004].) Figure 4.16 shows what one looks like, and the other place you will find one is as the connection between the antenna and a wireless receiver (for a mic) or transmitter (for personal monitors).

Figure 4.16 BNC connectors like the one here are used in making word clock connections as well as connecting antennae to wireless receivers. Image courtesy of Neutrik.

And there is our transition. Cables are not the only way to get a signal into the system. Sometimes you do it via radio waves. Welcome to the wonderful world of wireless.

5 The Wonderful World of Wireless

The previous chapter appeared to be about cables and connecters. But what it really comes down to is moving the signal from the source and into the system. There is another way to do this that does not involve cabling; it's called *wireless*. There are so many flavors of wireless guitar and bass packs, vocal mics, horn mics, and monitor packs sold at regular old music stores that it is easy to be fooled into thinking they are simple. This would be a mistake, because wireless is far from simple and is, in fact, one of the most common reasons for a show to go south.

Although it was once pretty clean and easy, the world of wireless has changed a lot in the past couple of years as things such as cell phones, wireless PDAs, and wireless laptops have become the norm. The other wildcard here is that depending on what part of the live audio world you end up working in, you may deal with wireless rarely if at all.

If you are in the backline business (everything onstage that is behind the PA—there is a whole chapter on this part of the biz later in the book), expect to deal with a lot of guitar wireless. If you don't do backline, you will likely not have to deal with it except for very rare occasions. If you are working for a regional sound company, you will have to deal with wireless mics and monitor systems regularly. But for every act that wants that equipment supplied as part of their rental and wants the sound company to deal with it, there is at least one other act that carries all of their own wireless, and there is a member of their crew who takes care of running it.

You need to know it not only for those acts that demand it, but also because it makes you a more valuable member of the crew. The guy who really "gets" wireless and can make it work and troubleshoot problems is never the first guy to go in a layoff situation. Lots of guys know how to mix and set up a system. Far fewer really understand wireless. If you want to make yourself indispensable, then learn this stuff forward and backward. Oh, and one last thing before we get into the nitty gritty: Even when the wireless is artist-provided, when it does not work, they will blame it on the sound company. Count on it.

Getting Unplugged

My first wireless came from W.A.S.P. guitarist Blackie Lawless, purchased in the mid-'80s right about the time that infamous metal band was starting to take off. My guess is that he was selling it because he got something better, but I didn't ask. Just standing in his living room stuffed with skulls, medieval weapons, torture gear, and horror movie books and paraphernalia was enough for a kid from the suburbs. I gave him the dough and got out as fast as I could. As I recall, that unit was from Nady, and I used it until I lost the screw-in antenna. Much like Nigel Tufnel in Spinal Tap (come on, you've got to remember the infamous gig at the Air Force base? If you don't, then homework for the night is to go out, find the DVD, rent it, and watch it), I endured dropouts, static blasts, and picking up other radio broadcasts in exchange for the freedom to move about the stage at will. Fortunately, the wireless of today bears only a passing resemblance to those early units—especially when it comes to price/performance ratio. Now, I use wireless for my guitars and for my personal monitors and generally have between three and six wireless mics onstage when I play with my own band.

Although my system is a mishmash of low-end pro and stuff that is just step or two above entry level, up until about a year ago, I never had a dropout or problem. I give most of the credit to these rules that I received from people on high who really understand wireless (also called RF for *radio frequency*). On the surface, wireless microphones, guitar packs, speakers, and earphones are extremely enticing for the performer who wants to cut the cord and enjoy total freedom of movement. But, as with all technologies, there are certain potential downsides that need to be respected. (Check out the "Ten Commandments of Wireless" sidebar for some important tips that can make life on stage much better for you and your audience.)

A Little History

Early wireless mics were long antenna VHF (*very high frequency*) types of units that were good at preventing dropouts in signal coverage but were clumsy in that they had large antenna aerials and limited frequencies of operation. Today's wireless mics and diversity receivers are mostly UHF (*ultra high frequency*) with a lot more frequency agility and high-quality audio to the point that (at least among high-end pro units) wired and wireless units have no discernable difference in audio quality. With some of the frequency crowding issues we are seeing, several manufacturers are looking anew at the largely fallow 900 MHz spectrum for their digital units.

Compression/Expansion

To make the wireless mic experience comparable to that of wired mics, a wide audio dynamic range is required. Typically, this is around 90 dB signal-to-noise ratio, like other live audio signal processing. Unfortunately, the frequency modulated (FM) channels must be as tight or tighter than normal FM radio channels, and that means about 50 dB signal-to-noise ratio. To obtain the extra 30 dB or more of dynamic range, an audio compressor circuit is employed to squeeze 90 dB down to 50 dB or less before frequency modulating the RF signal for the antenna. On the receiver side, once the RF switch has routed the strongest signal to the FM demodulator, the resulting received audio is expanded back to 90 dB and ready to by sent onward to the audio output jacks. While the diversity receiver has plenty of room and plenty of power to do a high-quality job of audio expansion, the corresponding compressor on the microphone has very little battery power and very little circuit board space to do the same quality signal processing. This is why many manufacturers will have different qualities of audio compression/expansion along with RF frequency flexibility for you to choose from. Obviously, a $1,000 wireless system is expected to have flawless audio quality compared to a $300 system.

Limited Power

Most wireless mics have limited power output capability due to Federal Communications Commission (FCC) regulations for unlicensed operation in the UHF bands. Typically, this is 50 milliwatts of antenna power or less. While radio frequencies diminish exponentially from the antenna, like audio signals from speakers, the modern diversity receivers can pick up very weak RF signals from tens to hundreds of feet away from the microphone. Having said this, most UHF transmission works upon a straight path between the antennas, or "line of sight" communication. This means that a bunch of people or a masonry wall is not expected between your receiver and the wireless mic's antenna. Thus, the best location for wireless mic receivers is off to the side of the stage or onstage—not out at the mixing console in the back of the audience. Also, both the wireless mic antenna and the receiver antennae are mildly directional, with minimum signals coming out the ends of the antennae. So do not point the antennae on the transmitter and receiver directly at each other, but let them point away so that the signals coming off the sides of antennae/aerials will allow those invisible RF waves to expand from antenna to antenna. Obviously, a wild mic-handling vocalist cannot control the transmit mic antenna orientation, but point the receiver diversity antennae upward and diagonally to catch as much signal as possible. And of course, obey the Ten Commandments of Wireless for your best wireless mic experience.

Ten Commandments of Wireless

(I) **Thou Shalt Have a Wired Backup.** Don't just have it with you, have it in place and ready to go. In other words, there should be a cable on the floor that you can be plug into your guitar at one end and your amp at the other in mid song if needed. True, the wireless units out there today are far better than in the past, but they can still fail. There is nothing like standing onstage in front of an audience, futilely trying to get sound out of a dead wireless connection.

(II) **Thou Shalt Do Thy Frequency Homework.** Know what frequency you are transmitting on and know how to change it if you need to. I recently heard about a band that almost got fired from a really good gig because the guitarist's wireless was on the same channel as the dude in the metal band next door. You never know until you get there what the wireless-spectrum situation is at a venue. There are venues in Las Vegas where so much wireless is already being used that plugging yours in requires a presidential decree. If you know what range you can transmit in, it will make your job—and the sound guy's—much easier if you encounter interference.

(III) **Thou Shalt Not Insist On Using Thine Own Wireless.** This is a companion concept to the idea that no matter what mic you prefer or carry with you, be prepared to use the standard-issue dynamic that the venue provides. Plugging your "alien" instrument into their "finely tuned" system could cause a problem. This is more an issue with wireless mics and personal monitors than with guitar units, but if you have specified wireless PMs and mics, and the house has a unit of the same model or similar to what you usually use, use theirs.

(IV) **Thou Shalt Have Diversity.** Diversity means more than one signal. At first, that meant two receivers with two antennae and a microprocessor that monitored the input to each receiver and then chose the strongest one to send to the output. Building two units into the receiver was not cheap, so diversity was reserved for high-end pro gear. Then, someone figured out that they could accomplish the same thing by using two antennae with the processor picking the strongest signal and sending it on to a single receiver. This led to the terms *true diversity* (the receiver kind) and just *diversity,* which usually means antenna diversity. Technology changes quickly, and for a brief period in time this made a real difference, but, as processing power has increased exponentially and antenna technology has advanced, the ubiquitous antenna diversity we see today is at least as good as the true diversity of days gone by. The bottom line is that if you don't see two antennae, there is no diversity, so consider carefully whether it is a good investment. Note that this does not apply to wireless personal monitor units because they are "backwards" from other performance wireless. That is, the receiver, not the transmitter, is what is in your beltpack. I only know of one diversity PM system on the market right now.

(V) **Thou Shalt Not Commit the Sin of Parallel Antennae.** If you have a diversity receiver of any kind, and you set it up with the two antennae on a parallel plane, you are all but negating any advantage of that diversity. Two antennae positioned close together and parallel to each other will—with almost total certainty—pick up the exact same signal. But if you put the two antennae at any kind of angle to each other, you vastly increase the possibility that one will pick up stronger signal than the other. The farther apart the antennae are, the less difference this makes. But, because most MI (*musical instrument*)–grade wireless receivers are housed in half-rack units, the antennae are no more than eight inches apart, and in this case, eight inches isn't a lot. . . .

(VI) **Thou Shalt Understand Companding.** To properly transmit any musical performance requires a fairly large signal in terms of bandwidth. The problem is that the bandwidth used by performance wireless devices is fairly narrow. (Though there are some exceptions that are addressed elsewhere in this chapter, most performance wireless devices operate in the space between stations in the VHF and UHF television bands—a space that is crowded and getting more so by the day.) To use the least amount of bandwidth, the signal is compressed before it is transmitted and then expanded on the receiver end. This process of compressing and expanding is called *companding.* Why is this important? Because the better high-end pro wireless units now available do their companding so transparently that only the most golden of ears can really hear it. The downside is that those units are outside the budget of the typical entry-level shopper. Less expensive and older wireless units use companding that you *can* hear. Generally, a companded signal has less dynamic range than one sent over a standard cable, so many performers have adapted the use of that compressed dynamic range as part of their sound. This is especially true of guitar players, who use that wireless compression to add to the sustain of their solo sounds. (Some digital wireless units just hitting the market that do not require companding including the Line 6 X2 units. Check the archives at L2PNet.com if you missed it.)

(VII) **Thou Shalt Take Care When Mixing Systems.** A frequency is a frequency, and systems of different "flavors" should work together just fine. In the real world, however, (barring any endorsement deals) you will almost always find that all the wireless in a rack is made by the same company. There is good reason for this. Most companies—especially at the pro end—make software tools that make setting up multiple wireless systems much less of a chore. They also allow the engineer to monitor things such as RF, audio level, and even battery condition from the front of house or monitor position. It can save the show if the engineer can tell you that your battery is dangerously low so you can change it between songs *before* it dies.

(VIII) **Thou Shalt Practice Proper Transmitter/Receiver Placement.** To avoid interfering with things such as TV stations, wireless units are very low power. This makes locating the transmitter and receiver in a direct "line of sight" very important. Beware of anything that can block the wireless signal, including your own body. Consider placing the receiver on the floor in front of you, rather than in a rack behind you, if possible.

(IX) **Thou Shalt Not Lose Track of the Mute Button.** This applies primarily to vocalists who insist on taking their wireless mics out into the audience (by definition, in front of the PA speakers). Unless the system has been thoroughly rung out to allow for such expeditions, this opens the door for big-time feedback. If you need to take the mic out into the audience, know how to mute that mic the second it starts to feed back.

(X) **Thou Shalt Carry Extra Batteries.** This goes without saying, but I'm saying it anyway. Have spares where you can readily access them and be able to change them out in the dark. While we're on the obvious, discard your old batteries ASAP (unless they're rechargeable), because they have this sneaky manner of finding their way back into your gig bag. D'oh!

These rules mostly apply to artists, so your best bet is to photocopy these pages and make sure the artists get them. The other thing to do is carry extra batteries yourself and charge $5 for the batteries you buy for $2. It can be a nice little source of side income.

And Now the Reality

This is complicated, geeky, and political all at the same time, but it is something you need to keep an eye on if you are using wireless anything on your gigs.

Most wireless units operate in the UHF frequency band. For those of us who remember TV before the days of cable and satellite, that means the channels above 13 on your old TV. Our wireless gear operates in the open spaces between those TV stations, which has worked pretty well most of the time for more than 40 years. But the times are a-changin', and they are changing on multiple fronts all at the same time. First, TV stations have—by governmental decree—moved from analog to digital signals. It is part of the whole HDTV thing, but like all things governmental, it also has a lot to do with money. Digital signals take much less bandwidth than analog, so portions of the VHF and UHF spectrums will be available for other uses, and the powers that be fully intend to tax and regulate them.

But the clearing of space that was expected by some in 2009 did not really happen. Until then, many broadcasters were transmitting both digital (for those with HDTVs) and analog signals at the same time, which has actually made the spectrum more crowded. Now that analog has been dropped, one would think there would be more space. That is a great theory, but broadcasters are looking to use that spectrum to offer additional premium services, so there's no new space for us.

Now, to make it really complicated, though we operate in the space between channels, that space does not really belong to us. In other words, we are kind of wireless squatters, so when a broadcaster that has paid big bucks for the right to use a certain part of the spectrum starts using the part that had lain fallow for years (and that we were using because no one else was), there is not much we can do about it.

This is why what we refer to as *frequency agility* is so important. You want any wireless you buy to have a lot of frequency options and not be limited to a small part of the overall spectrum. Believe it or not, it gets worse. Some very powerful, high-tech companies, including Microsoft, Google, Yahoo!, HP, and Dell, have petitioned the Federal Communications Commission to approve a new class of wireless consumer devices—likely portable Internet access devices. Because the wireless landscape theoretically opened up when the digital switch was thrown in 2009, and all of those analog broadcasts went away, they are arguing that the "white space" between channels will be wasted, and they want to make unlicensed consumer devices that operate in that part of the spectrum.

The bottom line is that very soon the sound of another broadcast being picked up by your wireless mic or guitar and coming through your system instead of the wicked solo you were expecting could move from an occasional annoyance to an everyday occurrence. This is a situation that will eventually require a technological solution, and every wireless manufacturer is working on some version of a solution right now. In the meantime, it would not hurt to let your elected representatives know that wireless gear is important to what you do and that making us go back to all-wired stages would put a real crimp in a lot of performances. The forces on the other side are huge, with towering piles of cash that they're not hesitant to spend. But politicians sometimes listen to voters, and enough calls and letters might buy us the time we need to develop a technology solution.

Dynamic or Condenser?

Back to the gear... Like wired vocal mics, wireless mics come in both flavors. You will only find a few examples of condensers for less than $1,000, because they tend to be quite a bit more expensive than their dynamic brethren. The same pros and cons that apply to wired vocal mics apply to wireless condensers versus dynamics. The biggest is quality of sound versus controllability onstage. Condensers have become more popular in live settings in direct proportion to the number of performers using personal monitors. The nature of a condenser mic is that it tends to have a more "open" pickup pattern and is more prone to feedback and bleed from other sources—there is a price to be paid for that more detailed and airy sound. The bottom line is that, wired or wireless, if you tend to mix acts on loud stages or use wedges for monitoring, you are probably better off with a dynamic.

The other issue is power. We had this bite us in the rear on my day gig recently. We bought a video rig for doing interviews and went all out and got wireless mics with it. But we went with clip-on mics (called *lavaliers*), not handheld, and that meant a beltpack like a wireless guitar rig would have. But—surprise—the mics are condensers, and the beltpacks do not provide phantom power, so we had to find a way to get power to the mics. In the case of a wireless vocal mic of the condenser variety, the issue is battery time—the batteries have to provide power to both the condenser mechanism and the transmitter. A dynamic only has to power the transmitter, so battery life is, in most cases, longer.

What's the Format?

Here we are really talking about physical form. Do you want a rack-mounted receiver or one designed to sit on a table or the floor that may be able to be racked with an adapter of some kind? As much as I hate it when people answer a question with another question, that's what I'm going to do here. How are you going to use it? If you mix three or four different bands, then you may want something more portable. If you work with just one act, then your receiver goes into a PA rack, and you're done. Keep in mind that more pro, higher-end models are only available in a rack version. If you need the good stuff and move between acts a lot, then invest in a small two- to four-space rack. A power module, your mic receiver, and, say, a personal monitor wireless transmitter, and that four-space rack is nearly full.

Auditioning

First, do your homework and check out what people in your area are using. Try to borrow or rent a couple of different models and use them on gigs before you make a decision. If you can't do that, then have the guy in the music store hook up several models and wire them all into the same PA. Set everything flat (no boosts or cuts in the channel EQ) and compare sound and response that way. *Do this with your vocalist.* A mic is a very personal and subjective decision. Comparing different mics in different PAs or in channels with different settings does not make for an apples-to-apples comparison.

After you have done the level-playing-field comparisons, start dialing in your sound on the models you liked best in the earlier comparisons. If you still end up with a couple of models you like, encourage your vocalist to start comparing how the mic feels in his or her hand. Look at how easy it is to change the batteries (as in, can you do it onstage in the dark?). How is it built? Remember, metal will always outlast plastic and take more abuse. If you are still undecided, look at the warranty, the reputation of the manufacturer, and the recommendations of the person doing the demo for you or—better—your friendly neighborhood sound-hound. If all else is equal, then it comes down to price. But don't rush it. A good wireless mic should last for several years of even tough gigging. It's a shame to buy a piece of gear on a whim and figure out a week later that you don't really like it or that it does not meet your needs. Be smart.

Time to Taste the Freedom

Some time back, I was doing a guitar sub gig with an '80s band out on California's central coast. I was using the gig as my road test for a number of universal-fit personal monitor earpieces. To keep from introducing an unknown factor (that would be a wireless monitor setup) into the equation, I opted to go with a wired beltpack. Knowing I would be anchored to a cable for the beltpack and remembering that the stage at this club is pretty small, I opted to leave the wireless guitar rig at home as well. By the third set of the first night, as I was trying to keep from tripping over the tangled wires coming from my guitar and to my PMs, I was reminded that going wireless is almost *always* worth it. If you are already a wireless user, you need no convincing as to why it is cool to cut the cable. If not, there is nothing I can say or write to convince you. It is just something you have to try in order to really understand why it is such a cool thing. So let's leave the "why" part out for now and take a look at the wonderful word of RF.

R What?

For those of you of the less technical persuasion, RF stands for *radio frequency* and is the term most often used to label all of the wireless devices in use in a typical musical performance. It is also seen by even many pretty technical folks as bordering on the world of black magic. If you *really* understand RF (and I only know a couple of techs who do), then you can get why the setup that worked one day is all of a sudden cutting in and out in the same venue with all of the same settings as before. To most of us, the best we can come up with is to shrug and say something helpful like, "Well, it must be Tuesday." Really. When you get into big productions that use dozens of channels of wireless, you will generally hire one of the handful of companies around who so get the RF thing that they have made it what they do. Companies such as Wireless First and ATK are the ones who get the nod for things such as Super Bowl halftime shows. And most of us are not sure whether we should approach these guys as fellow techs or as some kind of high priests with cosmic knowledge that we may never attain.

Now that we have made it clear that you will likely never really understand RF, let's look at some of the basics of wireless. If you are using wireless monitors, it is really easy to understand. What you have is the equivalent of a small, portable radio on your belt with headphones plugged in. But this radio only picks up one or, at most, a small range of frequencies that are not generally used for commercial broadcast. For your purposes here, the "radio station" is the transmitter that is connected to the mixer that is transmitting your mix via the airwaves to the radio on your hip. A wireless mic or instrument rig is the opposite—the radio station is the beltpack or is built into the handheld mic, and the radio is the receiver that plugs into the amp or sound system

The Magic Is in the Motion

Whether your artist is wearing/carrying the transmitter or the receiver makes a difference. On the monitor level it is easier because it is a traditional radio model—the transmitter is stationary while the receiver is in motion. Because the transmitter lives with the console and uses a constant power source (in other words, it plugs into the wall), it can transmit as strong a signal as the law allows, which is why a wireless monitor rig is less likely to suffer from weird dropouts, dead areas on the stage, and weak signals. A mic or instrument rig, on the other hand, uses a transmitter that not only moves, its output power is directly linked to its power source—a battery that loses power over time and eventually dies. This is a big part of why mics and wireless instrument rigs are more prone to dropouts and such, although things such as antennae placement can go a long way to helping that.

Now you know the difference between receivers and transmitters, so let's get down and dirty. There are a lot of wireless units out there in a huge range of prices. What gives, and are the expensive ones really better?

Your wireless stuff may seem like just another part of your musical gear, but remember that it is, in fact, a small radio station serving a single receiver, and as such, it falls under the authority of the Federal Communications Commission (the FCC).

It works like this: There is a limited amount of space available in any given frequency band (AM, FM, UHF, VHF, and so on) and lots of competing interests trying to get a piece of that pie (actually, it is known as *bandwidth*). So the FCC decides not only how big a slice of that bandwidth you can have, but also how much power you can use to transmit in said frequency range. For example, my favorite radio station in L.A. is Jack FM, which is at 93.1 on the FM dial. If I go to 93.1 in Las Vegas, I get a much less interesting classic rock station. (Jack is at 100.5 here.) So the FCC has granted two entities the right to transmit in the 93.1 part of the spectrum, but it limits the power of each so that they do not interfere with one another. Wireless devices for instruments and pro audio are always low-power affairs so they can operate in the same frequency range as, but not interfere with, the broadcast being offered by the person or company with enough money to hire a lobbyist.

On the compression side, we call it *companding* because the signal is compressed before transmission in order to cram all of it into the small slice allowed by the FCC and then expanded again on the other end. Compress plus expand equals compand—get it? The problem is that it is just not that simple.

If you want a real hearable test of this, then take a song from a CD you own and rip it to an MP3 file. Unless your system or ears (or both) are lousy, you will hear a difference between the two recordings. The MP3, in the process of compressing the data to make a smaller file, loses some sonic detail and dynamic range. Now take the MP3 and save it as a standard AIFF digital audio file, like on a CD, and listen. Most of the stuff that was lost in the MP3 remains lost in the "new" CD-format file.

But there are efforts to combat the effects of companding. There was a very good digital wireless system made by a company called X-Wire that worked like a big wireless modem for your guitar and sounded great. (Because digital data takes a lot less bandwidth to transmit than an analog signal, no companding was needed.) X-Wire was bought by Sennheiser, and the technology pretty much disappeared, though some other companies are looking to develop digital wireless products suitable for the pro audio and MI user. Lectrosonics makes an analog/digital hybrid that sounds so good you can use it with a measurement mic to sonically sample and then "tune" a system to the room in which it is used. And Shure has come out with something recently called Audio Reference Companding that uses new algorithms to combat the sonic effects of companding and does a hell of a good job with it. I recently used their newest wireless on a gig, and it sounded amazing. Yes, it is expensive, but in this case you get what you pay for.

Meat Absorbers

Some random thoughts on getting the best possible performance from your wireless gear:

- When it comes to wireless reception, the biggest impediment is distance. Always set up your wireless receivers as close to the performer as possible. The second biggest problem is the performer himself. The difference in power reaching the receiver when the transmitter is in the line of sight of the receiver and when the performer's body is between the transmitter and the receiver is about 30 dB. If you add a distance of, say, 75 feet (for example, if you put the receivers at the FOH mix position instead of at the side of the stage), you virtually guarantee dropouts.
- Diversity, in the wireless world, has nothing to do with being PC. It is a way of labeling units that can take two versions of the same signal and figure out which one is stronger and use that, switching inputs as signal strength changes. At first there was just diversity, and that meant two actual receivers in the box with one antenna serving each receiver and a microchip determining which signal got sent to the output. As the pressure to drop prices got more intense, designers figured out how to have the switching occur after the antenna and before the receiver. This meant one receiver in the box, which meant more affordable boxes. Although this antenna diversity was once seen as an inferior way of doing things, the technology has matured to the point where true diversity is almost never found in even pro-level gear.
- Finally, antennae placement. Remember that the transmitter is omni-directional. In other words, its already paltry FCC-imposed power goes out equally in all directions, so the antennae on the receiver pick up just a small fraction of the total power. If you have two antennae and you have them parallel, then they are likely both receiving just about the same signal, and the diversity switching doesn't really do anything. By putting them at differing angles, you have a better shot at one picking up a stronger signal than the other and that changing as the performer's position relative to the receiver changes.

Thanks to *FOH* magazine's technical editor, the late Mark Amundson, for making this simple enough that even I could at least pretend to understand it.

6 Snakes and Splits

So far, everything we have looked at has been all about taking sound (or acoustic energy), converting it into a form the system can use, and transporting it into the system. So you would think that now we'd get to start looking at the big parts of the system—and if we were living in the '60s, that might be true. But as technology has advanced, so have the demands of those outside of the audience.

What do we mean by "outside of the audience?" Let's take a look at a state-of-the-art venue and what looks like a good old-fashioned rock tour for examples.

A Pirate Looks at 40 Channels

Rich Davis and Billy Szocska have been the sound team for Jimmy Buffett for years. Buffett may not be your cup o' rum, and his music might lead you to believe that the production is simple. That would not be a good assumption. You see, there is a lot going on besides what the live audience is hearing. Monitor engineer Szocska makes sure the band and the boss are happy, while Davis mixes for as many as four audiences at once. There is the audience mix, a mix for the video team, a mix for the live radio broadcasts, plus at least 16 channels of recording, all off a classic analog Midas XL4 console.

And the ability to run that kind of system and that many mixes is a big part of why Jimmy Buffett is one of the highest-grossing touring artists in the world. You may not hear it talked about a lot, but those outputs are part of a business model that has made Buffett rich and kept his crew and sound company of many years (Sound Image, based near San Diego, California) busy.

Meanwhile, Back in the Desert...

The Maloof brothers make a whole bunch of money as owners of the Palms Resort and Casino in Las Vegas, the place where celebs and assorted other rich and beautiful people come to party and spend, spend, spend. When the Maloofs decided to put in a world-class performance venue, they did not mess around.

When an act performs at the Pearl, located inside the Palms, there are multiple audience mixes because there are areas for "regular" people plus Vegas-style VIP areas. (There is even a send to the restrooms and the bar areas.) There is a split to the monitor board for another mix. There is a split that runs from the stage, up 17 floors, to a world-class recording studio, which is tied right into a VIP suite in the hotel. There is another split available for sending to a broadcast truck.

As shows become more complicated and expectations grow ever higher, the job of the sound engineer has become much like that of a network engineer. There is one person at the monitor console and one person at the FOH console, but there are many more than that on a typical sound crew. And a big part of the job is getting sound from the stage to where it needs to be in the venue.

Splitting Sound

On all but the simplest gigs, the sources at the stage will go not to a console, but to some kind of splitter. These are usually referred to by a number—a two-way split or a three- or four-way split. In the days when all consoles were analog, it was not unusual for a monitor-specific console to include a split, but it is more common for the split to be separate from the console (although the difference between analog and digital here is huge).

Check out the pictures in Figure 6.1. This is a smaller analog two-way split. The first two rows of XLR connectors (all female) take an input directly from a sound source, such as a mic, or from a subsnake. The signal is then split, with one signal going to the male XLRs labeled Direct Outs and the second to the blue connectors labeled Iso Outputs. *Iso* refers to the fact that the last group of outputs is "isolated" from the inputs and direct outs by use of a transformer. Each output also has a Ground Lift switch, which is useful for eliminating the 60-cycle hum common in badly grounded electrical systems.

Figure 6.1 A splitter. Image courtesy of Whirlwind.

The snake allows us to move the audio from the stage to the front-of-house position over a single cable. The snake may terminate in another box of connectors and need short cables to get the signal into the console. Snakes generally end in what is referred to as a *fan*—several feet of the conductors inside the sheath on the snake exposed and terminating in an XLR or a TRS connector. Try to imagine lifting a bundle of 56 mic cables and wrapping it. It's heavy, dirty work, and if you are lucky, it is where you will start in the sound biz.

Figure 6.2 shows a small 24×8 snake. This refers to 24 sends (from the stage to the console) and 8 returns (from the console back to the stage). When looking at a snake, the first number is always the number of sends, and the second is always the number of returns. In a snake of this size, it is possible that the cable assembly is permanently attached to the box. But more likely, it attaches via a screw-on multiple-pin connector generally referred to as a *mult*.

Figure 6.2 A 24×8 snake made by Whirlwind. Image courtesy of Whirlwind.

Yes, snakes and splits are becoming increasingly digital and, by extension, smaller and lighter. But trust me, there will be analog copper snakes out there being used for a long time.

Ones and Zeros

The basic concept of the split and snake remains the same in the digital domain, but it looks a lot different. It has the same XLR connections on the stage end and does use another box at the console end. (The first time I used one of these, I quickly realized that I needed a bunch of three-foot-long cable to go from the box to the console, and not the

20-footers I usually carry, as the cables piled up behind the console in what had to look like some kind of viper's nest.)

The difference lies in the cable. With digital audio transport, an entire 100-foot length of snake takes up less space and weighs less than maybe six feet of the analog snake. That's because instead of individual conductors for each signal, a digital snake converts the signal from analog to digital at the boxes and transmits it all together over either fiber optic or standard Cat-5 or Cat-6 cable. The box at the other end converts the signal back to analog where it goes into the console—that is, if the console is analog. On a digital console, the connection will be made directly into the console or into its processing unit, depending on the brand and style of digital system.

But we are getting way ahead of ourselves. For now, just note that you will usually need multiple versions of the input signals sent to various places in the venue, and you use a splitter to make that happen. The cable that carries all of those signals is called a *snake*, and it can be a big, heavy bundle of copper or a single piece of fiber-optic conductor.

Oh, one last thing: subsnakes. It is not unusual to find several smaller versions of the snake around the stage that can carry anywhere from 8 to 16 signals. These are used just to keep the stage mostly cable-free. Instead of 12 individual cables going from the drum kit to the split, you plug each mic into an input on a stage box, and they all travel on a smaller snake to the split. It's just neater that way.

7 It's Not the Car, It's the Driver

The mixer (also referred to as the *console* or *desk*) is almost literally the heart of the system. Everything comes in, gets swished around, and then gets pumped back out as a finished product. The only thing left is to deliver that product to the audience.

Today's mixing consoles bear little resemblance to the gear we were using when I got started playing in bands and running sound in the mid '70s. But somehow, as consoles got larger, with more channels and more EQ and processing, and finally entered the digital domain, where they could do anything the most sophisticated recording console could do, sound at events both large and small got worse, not better. And strangely enough, the exceptions—that is, the good-sounding shows—are often mixed by the same people who were kludging together homemade systems in the earliest days of the performance audio business. They may be driving the most sophisticated system out there, but they only use the tools they need.

This chapter has nothing to do with gear or technology, yet it is very likely the most important part of the book. Harsh, but true...

A few years back, I was fortunate enough to say yes when a friend asked me to come to a small town in Utah to help with a festival gig. He had asked another friend, who also said yes. No one at the small high school where the festival was based had any idea that the guy with the Aussie accent working with the orchestra in the main auditorium was a performance audio legend. Howard Page, now the senior director of engineering for Clair Global, has mixed acts as diverse as Van Halen, James Taylor, and Mariah Carey. Here is what he had to say about the current state of performance audio.

> After being involved in live sound engineering for so long, I am very, very sad to see the way it has all evolved in the last few years. When did the kick drum become the lead singer? Show after show, regardless of the style of music, ends up being just a solid wall of badly mixed, way too loud, over the top, low-end-heavy noise. I have tried to help and nurture so many young guys over the years to understand what mixing live shows is all about, and my often-repeated sermon is to make it sound as close as possible to the recorded material by the artist. If some artists ever came out front at their shows and listened, I'm sure they would be horrified at how their performance is being brutalized. True, lately, some artists set out to use the sound

system to deliberately beat up the audience, but those shows are way beyond any help.

It comes down to really understanding your role as an audio provider. We refer to what we do by many names: Live sound, live audio, concert sound, and performance audio are a few. But I wish we could all get back to the term that really describes our job and function: sound reinforcement.

Our job is to not be noticed. We are there to help the performer communicate with the audience. It means giving the performer the means to convey their artistic intent and emotional vision past the area where they can do so unassisted. It is never about how cool your gear is or how loud you can make it or how bitchin' your kick sound is. We should be invisible and not affect the content of the performance in any way except to spread it further. Anything else should be considered as a failed gig.

Drew Daniels is an electro-acoustical consultant, studio musician, recording engineer and producer, and audio technology educator based in Los Angeles. His sound reinforcement experience includes stints with Teac, Fender, JBL, and Disney, where he filed no fewer than five patents. On his website, he lists four reasons for lousy sound at live events, and this is the best, most concise list I have ever seen, so I am stealing it. (Drew used the above quote from Howard Page, which originally ran in *FOH* magazine, which I edit, so I figure we're even.... Thanks, Drew.)

The List

1. Inadequate technical education
2. Hostility between sound providers and artists
3. Inadequate music education
4. Inappropriate gear or gear being used inappropriately

This book is, hopefully, a beginning point for taking care of #1 and, with a bit of luck, using the knowledge you glean here and from other sources, you can avoid the pitfalls of #4. Though it is really outside the scope of this book, we are going to spend a little bit of time on #2 and #3.

Hostility

I once talked with a good friend, a musician who spent many years touring with a well-known country act. I don't remember how it came up, but we got onto the subject of stage volume—the bane of sound providers at many gigs. When I said that keeping stage volume under control allowed the sound guy to make it sound better in the house, he replied, "Soundmen are the enemy." And unfortunately, this kind of attitude is rampant with bands. I have personally dealt with acts that were so loud onstage that I could not get the vocal up above the guitars, and the only thing still in the system was the vocal. I have had bands, when I asked them to keep stage volume to a minimum in order to get a good sound for the audience, tell me, "Drop it. I play loud." I have seen entire local concert series cancelled because one band was too loud and refused to turn down.

The act and the sound provider should be a team, so why the hostility? Many reasons. First, we, as pro audio providers, need to check our egos at the venue door. Remember, it is not about us; it is about the performer. Many artists have never experienced a situation in which the sound guy knew and practiced that concept. But I promise you that every time I have worked with an artist and have been able to communicate that I understand my role and am only there to make them sound good, I have gotten co-operation. Every time. There have been times when the level of distrust and hostility was such that I could not communicate that effectively, but I always try.

What is the cause of the hostility? 'Tis all opinion, so take it for what it's worth, but there are too many sound providers who started out as musicians and who are still carrying a chip on their shoulder about not making it, and they take it out on the acts they work with. This is very common in local and regional clubs with a house sound guy. Take a look in the mirror and make sure this isn't you.

Incompetence

The next reason is just flat incompetence among many house crews. I have seen some great house crews in my time. I have also, as a performer, had to deal with people who had no business behind any kind of sound console, had no idea how to operate the gear, and really didn't care. Too many venues don't put enough emphasis on their own sound. They hire unqualified people just because they will work cheaply. If you were a performer on tour and had to endure a string of such venues and crews, you would have your back up, too.

Sometimes the hostility stems purely from the fact that the artist is an egotistical jerk. But guess what? Even if that is the case, your job is to make him or her sound great. Refuse the gig the next time they come to town, but if you are there, then you need to do your job and actually care about how it sounds.

Education

The next item is inadequate musical education. This is a tough one. Many of the best sound engineers I know have no formal musical education, and they approach audio as a mix of art and science. Others are very accomplished musicians and know enough of the science to do the job but really approach it as almost another member of the band with the sound system as their instrument. (I know this seems to directly contradict the earlier statement about being "invisible," but the best way I can explain it is a quote from Tom Johnston, one of the founders of the Doobie Brothers. In an interview many years ago, he talked about the playing of his bandmate, Patrick Simmons, saying that few people understood how vital he was to the band's sound, and you often did not realize he was even playing until he stopped and the song fell apart. Try to apply that to live sound, and you'll "get" the artistic approach to sound reinforcement.)

The most important thing you can do with regard to music education is to familiarize yourself with the artist's music to the greatest degree possible *before* the gig. If you are working in a club that books four bands a night, most of whom you meet for the first time at sound check, then there is not a lot you can do. But being familiar with and understanding typical song structure in a given genre will help. Listening to lots of different kinds of music and having a solid understanding of how a great jazz band sounds versus the vibe of a great rock band is huge.

It comes down to knowing enough about different musical genres to know what is appropriate. Even that band you meet at sound check is not a total loss. Ask them who they think they sound like and what kind of music they listen to. This will give you at least an idea of the direction to start in.

Okay, lecture over. Just remember that it is about the music, not the gear or your ego. Now let's look at that console . . .

8 The Channel Strip

Okay, intrepid audio novices. When we last met, we had gotten through the first parts of the signal chain going from the source converting it via a transducer (mic) or direct box and getting it to the mixing console via either cables or wireless. So, we are at the console . . . now what?

Let's start with the inputs. (By the way, our examples will be analog mixers, as hopelessly old-fashioned as that may seem. I firmly believe that a solid foundation in analog is easer to build on than learning everything digital and then getting to a gig and finding out that you will be using a 24-channel, 20-year-old analog desk. I have seen young engineers faint in such circumstances and, yes, I took the time to ROTFL. These are located on the back of most pro desks, as you see in Figure 8.1.

Figure 8.1 I/O panel of a Mackie Onyx 1640i. Image courtesy of LOUD Technologies.

And sometimes they're on the top, as you see in Figure 8.2.

Or, they're on the back in the case of some units that are meant to be rack-mounted.

Inputs come in two flavors: XLR for mic inputs and usually 1/4-inch for line-level signals. (More on this in a minute.) The 1/4-inch can be either balanced or unbalanced. Again, most pro boards will always be balanced. I explained the difference in the chapter about cables, so I won't get into it here except to say that balanced is always better because it means a cleaner signal and less noise. On some boards you will also find RCA (phono)–style inputs (like the ones on your home stereo) for hooking up

Figure 8.2 I/O panel of an Allen & Heath ZED R16. Image courtesy of Allen & Heath.

things such as CD players without taking up a pair of "regular" channels. There are other jacks back there that we will get to later (inserts and outputs).

Now that we have the signal in the board, let's take a look at what happens first. It is not completely accurate, but the easiest way to look at what happens from the time a signal enters the console until it exits is to think of it like a series of pipes and water flowing through them. If the input is the source of the water, the first knob at the top is like a valve that determines how much water gets in to begin with.

That first knob is marked Trim, Pad, or sometimes Input. Note that the first control is *not* the fader or volume control on the bottom. That is often the end of the chain for that individual channel before it is fed to the master section. As I said before, that valve analogy is not exactly accurate, because it is at this point that a low-level mic signal gets boosted to a level the board can use, and this is where the magical mic preamps live.

That may seem like a bit of an overstatement, but sound guys will come close to fist-fights over mic preamps, and an entire product category has emerged as high-end mic pre makers have started selling a lot of external preamps to make up for the lack of "warmth" in many digital boards. This is really an argument over what a preamp is supposed to do. One school of thought says that all a preamp should do is boost the signal, while others actually change the character of the sound more like an equalizer might. So what some people hear as brittle or lacking warmth is often just a case of what comes in going out unaltered.

Part of the reason this whole preamp war is heating up is where the actual preamps are located in a digital system. Remember, in many high-end digital mixing systems, the console itself does not pass any actual audio at all. All of the "mixing" and processing happens in outboard units, and the console itself is just a control surface—kind of like a big, complicated mouse with lots of knobs and buttons.

In many of these systems—as well as most digital snakes—the mic preamps reside in a "stage box," or what we would call the split in the analog world. In other words, all of the mic signals get boosted right away, so that all signals hit the A/D (analog-to-digital) converters at the same line level. Now, this may make sense in the computer world (and remember that a digital mixing system is nothing more than a very specialized computer), but if I am doing what it takes to carry around a 600-pound analog console because I like the way the preamps color the overall sound, I am not going to be happy bypassing those preamps in favor of the super-clean ones in that digital snake. It is one of the big reasons why the most successful digital snakes to date are those that are a part of a larger digital mixing system.

Okay, back to our analog console. There are a few schools of thought about adjusting the input level. I am of the school that says you get a "full" signal at the beginning and then adjust levels later. The method for doing this depends on your console. If you have a button on the channel marked PFL (*pre-fader listen*), engage it; it does not really matter where the fader is. Bring up the trim control until the "clip" light comes on and then back it down to the point where it stops lighting up on the loudest parts of the performance. If you have meters, then look to get the signal in the –7 to –5 neighborhood.

Note that this is in Analog World. As I alluded to earlier, gain in the digital realm is approached much more conservatively just because digital distortion *never* sounds good.

Some boards—especially those ones you will encounter on club gigs and other smaller jobs—will not have that PFL button. In this case—in order to make sure that the relationship between input trim and output gain is correct—you will need to start with all of your channel faders at unity. This is usually an area along the travel of the fader marked with an arrow, a zero, or a shaded gray area.

The next section of the board is the EQ—a subject we will spend a chapter on, so hold tight for now. If there is a button to switch the EQ out of the circuit, keep it engaged until you have "level" on all of your inputs. If there is not a switch, make sure all of your EQ is zeroed out—in other words, nothing is being boosted or cut. Also, make sure that all of your auxiliary sends (auxes) are at zero, or "unity" if they are labeled that way. We'll look at what these sends do a little later.

At the beginning of this chapter, I mentioned something called *inserts*. (Sorry, outputs will have to wait, too.) An insert is like a detour that a signal takes between the mic pre and the rest of the channel. Processors such as compressors and pitch correctors go here—processing that you want the entire signal to get, and not just part of it, like what happens when you use an auxiliary send. (We'll get there; we'll get there!)

You need to use a special cable for an insert called—who would've guessed it—an *insert cable.*

It can be either 1/4-inch or XLR, but either way, an insert generally has both the send (the outgoing part of the detour) and the return (the incoming part of the detour) on the same jack. The cable is a Y with one connector on one end and two on the other that jack into the input and output of the device being inserted. On some boards (like most older Mackies), the insert doubles as a direct out for recording if you insert the connector halfway in.

A last note about insert cables—they change as you move up the food chain. On most club-quality mixers and even some pro models, you will use unbalanced single-point inserts, as described above. On higher-end consoles, you will move to dual-point inserts, which are balanced and require two cables.

By this point you should have mics and DIs set, the snake run, and signal running into the console, and you should have confirmed that there is an appropriate amount of signal coming into each channel. That's enough for now. In the next chapter, we'll look at EQ and auxes.

9 Console Auxiliary Sends . . . or, What Do the Knobs in the Middle Do?

In my day gig editing a pro audio magazine, we sell T-shirts with the answers to the questions most often asked of sound engineers by audience members. The first answer on the list is, "Yes, I know what all the knobs and buttons do, and yes, it took a long time to learn." (Thanks to James Geddes for that one . . .)

So by this point you have a signal, and it is controllable via a channel on your console. If the console is analog so is the signal. If it is digital, the signal has been converted somewhere along the line. The weak mic signal has been amplified, and now we can manipulate it. And here is where we enter dangerous ground.

Remember some time back I brought up the fact that our job is really to reinforce what is happening onstage and make it loud enough for a larger group of people? Ideally, you should not have to touch the EQ at all, but it hardly ever works that way. The same voice can sound very different through two different mics or speaker systems. Even mics and systems made by the same manufacturer can have subtle differences that you need to make up for. But that is EQ, and it is the next chapter. For now, we are going to talk about auxiliary sends and returns.

The size and "level" of your console will largely determine how many auxes are available and where they appear in the signal chain. Aux sends come in two flavors—pre-fader and post-fader (and often auxes are switchable between pre and post), and this position determines what they are used for.

Made to Order

You know how when you are at a bar and you order a drink (be it with or without an adult component), the bartender pulls out a little gun-type thing from which he can dispense all kinds of liquid refreshment? Aux sends are kind of like that but backwards. To continue the plumbing analogy we started some installments back, if sound is like water, then the console is like a series of pipes and valves that determine what goes where and how much of it.

An aux is like a valve that sends some amount of the "water" off to another system. In some cases the "water" is returned to the main system after something has been added, and in other cases some of it is sent off to do work elsewhere. Whether that "water" returns to the main system is based on what happens to it while flowing through the other system.

How you set up the flow at different parts of the system is known as *gain structure* and is the source of many arguments in many bars between many audio folks. But there is a reason for the passion behind something that may seem minor: Good gain structure can be the single biggest difference between a good-sounding show and a bad-sounding one.

Pre-aux sends are usually used for crafting monitor mixes. (EQ generally comes before the pre-fader sends in the signal chain, but some consoles will allow you to switch the EQ in so that it affects those monitor mixes as well as the main output.) Monitor sends are before the fader because if a change is made to the house mix, you don't necessarily want the same change to happen in the monitors (a real issue with digital consoles that share a mic pre, but again, a subject for another time). Therefore, the sound goes off to the monitor system *before* the main fader so that the monitor mix can be crafted apart from the main mix.

Now, just to make sure there isn't too much confusion... Yes, we talked about splits, and in a concert-sized rig, one of those splits will go to a separate monitor system. But most smaller gigs—the kind you will be working with early on—will often require that you run the mains and a few monitor mixes from the same console. Those extra mixes are the product of the pre-fader aux sends.

A post-fader aux send is generally used to drive some kind of effect or sound processing. This allows you to determine how much signal gets that reverb on it, and a separate aux return or channel determines how much comes back. Most good mixers have aux return controls in the main output section, but most pro sound engineers will return the effect through an unused channel so they can easily determine how much of the effected sound gets back to the system and to the ears of the audience. This is, for example, a great way to easily kill the reverb on a singer when he or she is talking to the audience. Reverb sounds cool in judicious amounts while singing, but it almost always makes spoken communication very hard to understand.

Insert Here

One other kind of send and return is the channel insert. This is like an aux send, but you don't get a knob to determine how much sound gets sent. It all goes, and it all comes back. This is usually used for dynamics processing—compressors, limiters, and gates—but it can also be used with an outboard EQ or automatic feedback killer. These get "inserted" into the channels where they are needed and are not on an aux.

Your better consoles will have insert send and receive jacks for each channel—usually balanced XLR connections, but almost all mixers commonly found on small gigs, including clubs, houses of worship, and the like, have a single TRS 1/4-inch jack. To use the inserts, you will need a TRS send and receive cable. This is a Y cable with a single

TRS on one end and a pair of mono 1/4-inch jacks on the other end. The tip of the TRS sends the signal to one of the mono 1/4-inch jacks while the ring portion gets the signal back from the other mono connector. To further complicate things, most pro consoles have direct outs that send the raw signal from the input off to other places, usually a recording unit of some kind. But some MI consoles combine the inserts and direct outs. If you insert a TRS all the way into the jack, it is an insert. If you stick a mono jack into the halfway point of the first click, it acts as a direct out.

Depending on the size and level (that is, MI, pro, touring) of the mixer, the master section will be either in the middle (typical of most higher-end models) or on the right-hand side. Here you will find your master volume faders along with your aux returns and usually a master send control. There may also be a two-track in/out with RCA jacks and a separate mono fader for a center cluster (used in a LCR, or Left, Center, Right, mix) or to feed a subwoofer. If the mixer includes subgroups or VCAs (*voltage-controlled amplifiers*), those faders will be placed here as well, and perhaps some rudimentary EQ.

Strange as it may sound, we are going to cover what seems like the simplest part—the master faders—in a later chapter because it means dealing with gain structure, and that is the subject of a whole chapter all by itself.

Remember Tapes?

Those RCA two-track jacks are the easiest to get out of the way first. The two-track out is a mirror of whatever is coming in to the master fader, and it allows for the easy connection of a recording device for making board tapes. If you have a two-track in, it allows you to insert a playback device (such as a CD player) without burning up a couple of input channels. In truth, it is something of a throwback to the days when we all had a cassette tape machine in the rack. These days, playback is more likely to be an iPod, and recording is done on a laptop. Some manufacturers now include USB or FireWire outs to feed that laptop directly. Also, I would not be surprised if, by the time you are reading this, someone puts an iPod dock connection right on the console as well.

A couple of two-track "extras" can include a separate gain control labeled something like "two-track to Aux 3," which allows you to send the playback to just the monitors if you want to get something like a pitch reference or click track to the band but not have it in the mains. There may also be a switch that says something like "two-track to Channels 15-16," which sends the two-track signal to a couple of input channels so you can use the EQ and aux sends from the channel if you need to.

Okay, now the next easy part. If your mixer has a small graphic EQ in the master section, it is most likely pretty useless, although there are a couple now available with digital 31-band EQs at the master. Most likely, you will have a seven- or nine-band graphic. If there is a switch to bypass these, use it. If not, just flatten it out and ignore it. An EQ that small is going to boost or cut things that you don't want to touch as a side effect of anything you *do* want to change. The bands are so wide that it is akin to doing surgery with an axe when you really need a scalpel.

Now Boarding Group A . . .

If the console has subgroups, here is where the master for those groups will reside. You assign like inputs (say, drums or horns in a section or backing vocals) to a group. You can then adjust the level of the entire group of inputs without adjusting individual channels.

VCAs serve the same purpose, but there is an important difference. In a subgroup, the audio signal itself routes through the group faders so you may have a direct out or an insert. With a VCA, no audio is present at the faders. The faders just send control voltage that raises or lowers the output of a channel or group of channels—hence the term *VCA groups.*

The advantage to this approach is that it means one fewer set of components for the audio to travel through, and it can result in a cleaner sound. This disadvantage is the loss of those inserts that allow you to, say, put a single compressor on a horn section.

I Wanna Go Home

Your aux masters reside here as well. You should remember that there are sends to the auxes on individual channels. For a pre-fader (in other words, monitor) send, this is the master volume control for a specific monitor mix. For a post-fader (in other words, effects) send, it determines how big a signal is sent to the outboard processor. The amount of the effect you actually hear in the mix is controlled by the aux return. Turning up the return combines more of the wet (effected) signal with the dry input. If you have too much reverb in the mix, it likely means that the effect returns are dialed up too high.

There may be another control labeled something like "return to Aux 3," which puts the effect in the monitors as well. Be careful with this one. Even if the singer likes a lot of grease on the vocal, the more reverb in the wedge, the greater the chances are for feedback.

10 Equalization

In talking about the console, we have been concentrating on the individual channel strips, and that will continue here. But there are other places in the system where EQ will come in again. Just so you know...

Rule #1: Listen First

You know how when you go to dinner with friends, there is always someone who picks up the salt shaker and starts salting away before they have even tasted what is on the plate? I hate that.

I also hate it when I see an engineer start making EQ adjustments before there is sound in the system. Every system is different, and you have to listen to get the most out of it. The most important skill you can develop when it comes to running sound (at any level, from a small rehearsal room, to a club, to an arena) is to learn how to listen. With all of the new high-tech toys available, I find far too many sound guys who spend more time looking at laptops, touch screens, and processor menus than they do listening to the band. They are, in effect, trying to mix with their eyes. This doesn't work very well. It's important not to get too tied up in where the knobs are pointing. Adjusting the EQ based on what you are hearing is far more beneficial than making sure a particular frequency band is knocked down by 6 dB like the guy in some magazine says it should be.

I remember watching one sound guy, whose ears I admire, adjusting a system. He did not even look at the knob—in fact, his eyes were closed. He turned it until it sounded the way he wanted it to. That is a great approach.

EQ Bands and Types

A typical MI (*musical instrument*)–quality mixer will have anywhere from two to four bands of EQ on each input channel. Two is easy—one is high and the other is low—just like the bass and treble controls on your home stereo. As we add bands, we get into the midrange, and that is where things can start to get confusing.

Let's start with EQ types. First, you need to know whether you are looking at a true cut-and-boost filter or a simple roll off. With a roll off, all of the frequency content of a particular frequency band is present when the control is dialed all the way on. Dialing it back "rolls off" the content of that band. A true cut and boost is at zero—or *flat*—when the knob is at 12 o'clock. There is often a notch in the knob's rotation at that point called a *détente.* Dialing the knob up or down either boosts or cuts the content of that frequency band. Both types of EQs are centered at a specific frequency and have a specific width (how many adjacent frequencies they affect) called the *Q*. These frequency centers and filter widths are a huge part of what makes one mixer sound different from another.

The other kind of EQ or filter is called a *parametric* or *semi-parametric.* These are also referred to as *sweepable* and are usually found in the midrange. A good console—for me, anyway—will have four bands of EQ including two sweepable mids.

A fully parametric EQ consists of three adjustments. First is the center frequency, next is the amount of boost and/or cut applied to the band, and finally, the Q control that adjusts how wide the band actually is. An EQ that includes all three of these controls is referred to as *true* or *full parametric.* Most of the sweepable controls you will find on MI mixers will leave out the Q control (the width of the filter is fixed) and are properly referred to as *semi-parametric.*

There is another member of the EQ family that you need to know about. It's called a *high-pass filter.* This is not a knob, but a switch that allows frequencies above a set point to pass and steeply rolls off anything below that. (A low-pass filter does just the opposite.) High-pass filters are found on most pro consoles and can work wonders for cleaning up a mix. Just engage the filter for any source that does not have content below the point at which the filter is set (typically about 100 Hz). You'll find that this is most of your mix.

How Do I Use 'Em?

The first thing I do with any board is to zero it out by setting all of the channel faders, auxes, and EQ controls at their zero settings. Remember, on a true cut-and-boost EQ, the zero setting is usually at the 12 o'clock position. As you gain experience and get a feel for your system, your mics, and the players, you will find yourself making the same cuts pretty much all of the time (such as cutting at 120 Hz to take the mud out of a kick drum or cutting 1.25 kHz from a vocal mic). When you get to that point, it is tempting to just make those adjustments automatically before really listening to the system. In my world, that is just a bad idea. Start flat and listen before you start adjusting.

Rule #2: 'Tis Better to Cut Than to Boost

When it comes to EQ, it is *always* better to cut than it is to boost. Remember our plumbing analogy from Chapter 9? Well, assuming your main pipe is pretty full to start with (as it should be if your channel trim is set right), then adding EQ is like adding water to that pipe, which could overload it. In the audio world, that means distortion and maybe feedback. When in doubt cut, don't boost. So, how do you get more bass, for example? Try cutting everything except the lows and then boosting the overall signal a little bit to get the same effect as just boosting the bass.

Table 10.1 shows the frequency bands of various instruments. (This is a little like giving away the secret of the ages for sound guys. Keep this info close to the vest, lest it fall into the wrong hands. . . .) Knowing the frequency ranges of various instruments can make the job of EQing a lot easier. Keep this chart handy until you have it burned into your brain.

Table 10.1 Frequency Bands of Several Instruments

	Frequency Range (Hertz)		
Instrument	**Minimum**	**Nominal**	**Maximum**
Kick Drum	130	164	196
Floor Toms	220	-	440
Rack Toms	350	-	700
Snare Drum	650	784	1000
Cymbals	500	-	40k
Bass Guitar	31	-	500
Guitar	82	-	700
Piano	27	-	4186
B3 Organ	32	-	5920
Tenor Sax	110	-	587
Violin/Fiddle	196	-	2093
Blues Harps	196	-	2959
Baritone Voice	110	220	392
Tenor Voice	146	261	440
Alto Voice	196	392	698
Soprano Voice	261	523	1046

Tommy Rat is a legend in the live sound business and the survivor of literally thousands of shows—many of them of the punk variety, which can be very tough. He is also a true mentor to a generation of sound guys. And he has a great system for learning what a given frequency "sounds like."

> One of the ways I teach people to reference frequencies is to sit them down with a microphone, a high-powered speaker, and an EQ. If you do this, and I suggest that you do, you can turn up the volume until you get a controlled feedback. Once this is

achieved, you can utilize the equalizer to find the frequency that is feeding back. By positioning the microphone in different places, you will be able to create different feedback scenarios and hone your skills for recognizing tones and applying them to their corresponding numbers. EQing is basic math, and every tone has an assigned number.

It is all about "ear training." A good experienced engineer can almost instantly identify a frequency that is feeding back. Honestly, I am not very good at it, so maybe I appreciate the ability more than most.

If your housemates or neighbors are not excited about you blasting feedback to attain this knowledge, and if you have an iPhone, look for an app called Dog Whistler. It is made as a dog-training tool, but it generates frequencies and identifies them, which makes it a good sound-guy training tool as well. Whatever tool you use, really knowing your frequencies will make you much more valuable on any crew out there.

11 Other Channel Stuff

Depending on how "pro" the console is, there may be some other controls that you need to understand.

Phantom Power

Depending on the level of the board, phantom power may be switchable on a per-channel basis in two or more groups of channels, or a single switch may turn on phantom power for the whole board. Generally, it is controlled by a switch labeled +48.

As we discussed back in Chapter 3, condenser mics need power to work, and this power is supplied to the mic from the console. (Active direct boxes also need phantom power.) There are myriad issues with phantom power, mostly in the form of lack of standardization. Some boards don't put out a true 48 volts, and most mics can use less. On lower-end, less expensive mixers, you are likely to find the phantom power switchable in groups or globally.

You may have heard that using phantom power on a dynamic mic can damage it. Kinda sorta, but that's not really true. It can damage an unbalanced dynamic mic, but you are unlikely to run into one of those. Same with ribbon mics. But the truth is that pretty much any mic made in the past 30 years should be fine. But more things than just mics go into your system.

Anything with a line-level output *can* be damaged by phantom power, and that includes keyboards, drum machines, some bass amp direct outputs, and especially consumer devices, such as CD players. Phantom power is only sent to the mic input, so using the line input should negate any damage issues. Problems arise when someone tries to get a 1/4-inch (almost always line or instrument level) into an XLR mic input. Why? Well, if the interface to the console is via a stage box and snake, it is an easy leap to use an adapter to change that 1/4-inch connector into an XLR and run it right into the stage box, right? Wait, what's that noise? And what's burning? And... well, you get the idea.

This is where a good direct box comes into play. The DI takes a line-level signal and takes it down to an appropriate level for the mic input. It matches impedance and converts the line-level unbalanced signal to a balanced signal that can run much greater distances without noise or RF (radio) interference. It also uses a transformer to isolate the voltage coming up the line and keep it from reaching the outputs of your line-level device. There are a few companies making specialized direct boxes for taking an 1/8-inch stereo mini jack (such as the headphone out on your iPod), converting it into two separate signals (left and right channels), and sending them on their merry way over an XLR connection.

Most direct boxes you will use are of the passive or unpowered variety. But for some applications, an active (powered) DI is called for. The active DI gets its power from a battery or from the console's phantom power. The big difference between the two is that the input impedance of the active DI is much higher than that of the passive DI, and that higher input impedance preserves all of the upper-register harmonics that a passive DI can choke out of a signal. This makes them especially good for things such as acoustic guitars, where you want that top-end "sheen" to the sound.

Mute and Solo

It's pretty straightforward—mute stops the signal from a channel from reaching the output. Solo is also referred to as *solo in place*, and you need to be careful. It mutes any channel that does not have a Solo button pressed. This can be useful during sound checks and line checks to isolate a single input, but if you hit it during the show, the audience will get treated to not the sound of a full band but, say, a really cool bass solo.

If you have buttons labeled either AFL or PFL, they are a lot more useful. AFL stands for *After Fader Listen* and will isolate the signal after the fader so you hear exactly what is going to the main section of the console. PFL stands for *Pre Fader Listen*, and it lets you hear what is happening without regard to the actual fader level. The difference here is that in most cases, the AFL or PFL will have no effect on what is hitting the master section of the board. Rather, it is routed to the headphone output, which allows you to listen to the isolated signal over the headphones. This can be very useful for adjusting EQ or effects on a single input during a show.

One other thing: Many mixers feature direct channel outputs most often used for recording. Increasingly—especially in midsized mixers—these outputs are digital over USB, FireWire, or ADAT Lightpipe. When those kinds of outputs are available, there may be some kind of button or switch that lets you determine at what point in the signal chain the direct out sends—pre- or post-EQ, pre- or post-fader, and so on.

Finally, the control that is (in my humble opinion) the most abused, misunderstood, and ignored on the strip is the pan control. Too many sound operators are under the impression that because it is really only possible to get a true stereo sound in certain parts of the room, they should just run in mono. Likewise, people playing stereo instruments have a bad tendency to try to give you a mono signal. All of this is a mistake.

First, let's address the instrument issue. The sounds programmed into a stereo keyboard, guitar processor, or whatever are *made* to be heard in stereo. Yes, they may have a "mono" out. (Usually either the left or the right output will also be labeled *mono*, and when there is nothing jacked into the other output, you will get some kind of "summed mono" signal that should have all of the audio info on one output.) But in reality, it just doesn't work that way—especially with sounds that include some kind of modulation effects. You can't get the swirling, motion-filled sound of a stereo chorus or even a Leslie speaker in mono. Not gonna happen... It is like you are starting out with one hand tied behind your back. Yes, if you don't have stereo channels on your board, it means burning two channels for one instrument. That is why you get the most inputs you can afford when buying or specifying a console.

On the whole "sweet spot" issue: Yes, it's a problem, and yes, only a small part of the audience will get a "true" stereo experience. Big acts have gone to major expense to try to alleviate this.

One approach is to do a standard left and right at the front of the stage, and as the system moves to stage left (your right when you're facing the stage) there is another stereo set, but this one is reversed, with the right channel on the left and vice versa. Why? People sitting closer to the corner of the stage might hear sound from both the main speakers at the front of the stage and the ones at the side. By repeating the right channel, that part of the image stays coherent as it wraps around the stage.

Using this approach, the image is flipped on the sides of the stage, but most of the audience gets some kind of stereo sound. However, this is expensive and requires serious system design and acoustic chops that are way past the scope of this book. So why should you bother with stereo signals and panning even on a simple system? It is all about opening up the center of the sound image for the important stuff, such as the lead vocal. It is not about extreme panning, just moving things around a little bit.

This is very much a matter of opinion, and there will be sound guys who will violently disagree, but on a typical gig I tend to pan things a little bit based on where the player is onstage to keep some kind of integrity to the overall sound image. Take a look at the stage and try to move your pan knobs so that they basically represent the direction from which the source sound is coming. If the lead guitar is standing to the right of the lead vocalist, pan that channel a bit to the right. Pan your toms so that they move across the soundstage as they are played. Ditto drum overheads, but not hard panned—just somewhat off center. Stereo inputs such as a keyboard or stereo guitar processor get panned hard left and hard right. So, what is straight up the middle? Lead vocal (or primary instrument in the case of a non-vocal act), kick drum, and maybe snare. Everything else is panned just a little bit. Again, it opens up the middle for the "money" channel.

Another way that system designers and venues make sure the vocal is really out front is to implement an LCR (left, center, right) system. This format allows you to place supporting instruments into the left and right outputs and reserve the center cluster for your money channels.

12 The Master Section

So you have all of your inputs in all of their individual channels at the right level, EQ'd as needed and being sent where they need to go. So you're finished, right? Not quite. Next comes the master section, which can be thought of as a channel all on its own. In other words, it is not just those master level faders that you need to worry about.

Aux Returns

Any aux send that has the potential to run post-fader will generally have a return control in the master section because they have the potential to be used for effects. These returns are often just level controls and determine how much of the signal that was sent to the effect gets returned to the master section. In the case of, say, a reverb, the return is the overall reverb level control. The channel sends determine how much of the dry signal goes to the reverb. All of those signals hit the reverb and are processed according to the settings on the reverb unit. The output of the reverb goes to the return, and the setting of the return determines how much of the effected signal goes to the *master*, not back to the individual channel. If you have an effect that you want to send and return directly to a channel, you use the channel insert, not the aux send and return.

Most mixers will, given sufficient channels, bypass the effect returns and send the effected signal back to another individual channel. There are several advantages to this approach. The effected signal is another input that can be EQ'd and so on. Crucially, the effected signal can be muted when the person at the mic is speaking between songs. That killer reverb plug-in may sound great when the person is singing, but when he is speaking, it tends to turn the voice into incomprehensible mush.

We have already established that your early gigs—at least the ones on which you are allowed near the console—will not be with big touring acts. They will be with local acts, maybe regional bands in clubs and such. We need to be reminded of that now because on those kinds of gigs, you will rarely have someone else to mix monitors from a separate console, and you will be providing as many monitor mixes as you have pre-aux sends for on your console. Instead of a return for these sends, you will find a master send level that determines how much total signal gets sent to those monitors. Sometimes you will get a switch or, if you are lucky, a fader labeled something like Aux 1 to Aux 3. This allows you to put a little "grease" in the vocal monitor. But use this very carefully. If you are mixing for the house and monitors, you have your hands pretty full. That lead singer may like the way his or her voice sounds drenched in reverb, but if you are using stage wedges, just a little too much reverb in a wedge will put the feedback cycle into gear.

Pretty Lights

The VU meters of old have given way almost completely to LED-based meters that go from green to yellow to red as the signal they receive gets hotter. Any addition to the signal level—be it from an individual channel, an aux return, or the level of the main output—will affect this meter.

And this is one of the places where we have to think very differently depending on whether we are mixing analog or digital. On an analog mixer, you want to mix at as close to 0 or "unity" as possible. Given the dynamic nature of performance, this means that the meter will inevitably hit the red a bit from time to time, and that is acceptable. But when the mixer is digital, the standard is closer to –6 dB from unity as your average point to ensure that digital clipping never happens. Why the difference? Analog distortion is simply not as hard on the ears as digital is. In fact, in the analog world, what many musicians and engineers refer to as "warmth" in the system is really just a little bit of distortion, and to many ears it can make an overall mix sound better.

But digital distortion is a different beast, and no one who has heard it would refer to it as "warm." It is a very unpleasant sound that can be best likened to (sorry for the scatological reference) an electronic fart. Mixing at –6 dB makes this sound much more unlikely.

Other Miscellaneous Stuff

In this section, you will also find tools such as your headphone level, two-track input, and outputs, with level controls for each. You will also find a switch and level control labeled Talkback. Some semi-pro boards actually have a mic built right into the console next to this control, but usually you will need to plug a mic into the Talkback input to use this feature. And it can go a long way in saving your sanity and your voice. The band can always communicate with you by speaking into a mic. Talkback gives you the same ability. Assign it to an aux send, engage the switch, and speak into the mic, and the act should be able to hear you through the vocal monitors.

Anything that fosters communication between the act and those mixing the sound is a good thing, and some acts make it a standard part of every show. Jimmy Buffett sings through a normally routed vocal mic but wears a wireless lavalier mic in addition. That mic is routed only to the monitor engineer and specified band members. It allows Buffett to step away from the vocal mic and communicate directly with the monitor engineer. Billy Szocska, who has been doing Buffett's monitors for a few years, says that when he first started, he made the mistake of turning off that lav mic's output. "I did it once," he said in an interview and then repeated, "Once"—with an emphasis and body language that made it very clear that doing so had not made the boss happy. Billy will tell you that in many ways, that lav mic is the most important thing onstage. At least it is to the guy who signs the checks...

Splitting It Up

So far our talk of the master section has focused on bringing everything together, but there is another set of controls that kind of lies between the channels and the mains. These are either subgroups or VCAs (*voltage controlled amps*), and though both serve similar purposes, they do so in very different ways. And in fact you may actually have both on a console. And as long as we are talking about grouping channels—which is what both subgroups and VCAs do—we should talk about something we left out in the channel strip section. Remember we talked about the mute control? Some consoles will allow you to assign multiple channels to mute groups. Why? Let's say you are mixing an act with three background singers and a horn section. However, those voices and instruments are not used in every tune, and sometimes they may actually leave the stage. Assigning all of the backup vocal mics to a single mute group gives you the ability to mute all of those mics at once, making for less chance of missing one or neglecting to turn one back on. Okay, back to where we were before this little detour...

A subgroup is just what it sounds like: a feature that allows you to assign multiple inputs to a single group. So you can, say, put all of your drums in a single group, and once you have the relative levels right, you can raise or lower the level of the entire drum kit without screwing up that relative balance. For most purposes subgroups will work just fine in a live sound setting, but there are exceptions. Let's look at those backup singers again. Suppose you want to fade them out of the mix. Grab the subgroup fader and bring it down slowly, and it all fades out, right? Not quite. Remember we talked about post-fader aux sends used for effects? Well, unless you have multiple reverb units available and are running a dedicated insert on the subgroup, then fading out the subgroup will fade the primary signal, but because it only controls the main outputs of the channels in the group, the reverb on the vocals will continue to flow into the main mix. This might be a cool effect, but it is likely not what you are looking for. Or, say you are running your subs on a post-fader aux (an increasingly popular practice that allows you to easily control the sub-bass level from the console); with a subgroup, changes made to a group will not affect the amount of sonic info going to the subs.

VCAs are a whole different animal. In fact, they actually start back at the channel strip. VCAs replace the potentiometers at the channel strip. Sort of... The fader is still a pot, but no audio passes through it. Instead, the signal goes to an amplifier, the output of which is determined by the voltage applied to the amp. So in a VCA console, the fader is actually controlling the voltage sent to the amp, which raises or lowers the level of that channel.

In turn, you can assign multiple channels to VCA groups, which serve the same function as subgroups but, again, do so without any audio signal actually passing through the fader. Instead, the voltages of the channel VCA and any VCA groups are summed. So if your channel is set at +3 dB, and the VCA group is set at –5, then the actual output of the channel is –2 dB from unity.

A little too geeky? Let's look at the practical differences. We already looked at the example of a fadeout using a subgroup. In the same scenario using VCAs, as the voltage from the VCA group is lowered, so is the voltage—or level—of the channels assigned to that group. This means that any post-fader aux is affected as well. Think of it like this: With subgroups, the channels assigned are combined into one signal, the level of which can be controlled via the subgroup fader. But with a VCA, because the group control is actually affecting the output voltage of the channels assigned to it, the result is the same as if you grabbed all the faders of all the channels assigned to the VCA group and moved them up or down by the same amount.

If it helps, visualize the VCA master actually moving all of the faders in the group. In effect, that is what is happening here.

And remember that in most cases, a channel can be assigned to more than one sub or VCA group. This can make your job a lot easier once you have things set up. For example, the drum inputs all go to the same group. But you can also route them to another subgroup that includes all of the instruments and no vocals. That way, you can easily raise or lower the level of the drums or turn the whole band down when they start to overwhelm the vocals.

Finally, your groups will likely not have their own EQ or aux sends, but there is probably a group insert jack on the back of the console where you can add an effect or other processor and have it appear only on that group.

13 Gain Structure

Before we move away from the console and on to the drive rack, amps, and speakers, I want to take one more look at proper gain structure. Steve LaCerra has his own record label, is a very good studio engineer, and has been the FOH engineer for classic rocker Blue Öyster Cult for more than a decade. I have been fortunate enough to have him write for me in several different forums, including the Live2Play Network, *Front of House*, and *GIG*. So here is his take—from someone who does this every day and deals with a new PA every day. There is a ton to learn here. Some of it will repeat ideas we have already touched on, but they are laid out here in a way that is eminently usable. Thanks, Steve.

An extremely important aspect of sound system use, gain structure is the red-headed stepchild of audio: It's often misunderstood and typically neglected until there's a problem. Let's try to demystify the idea of gain structure so that you can get on to making music with the best sound quality possible.

What Is It?

Gain structure refers to the manner in which signal levels are set in (and between) the various sections of an audio system. Sounds easy, right? Wrong. Turn a level control too high, and you'll have distorted audio. Set it too low, and you may find that your system doesn't play loud enough or that you can't get sufficient level to tape. Symptoms of poor gain structure include noise of the hissing type (as opposed to hum or buzzes), distortion, lack of headroom, and grossly mismatched readings between the meters on different devices used in your system. It may lead you to believe that a piece of gear is malfunctioning or that you have a bad cable in the chain. When gain structure is set correctly, you'll get every last dB out of your PA, you'll record cleaner tracks, and all of a sudden your digital processors will have a better signal-to-noise ratio.

ABC's of the Signal Chain

One of the most basic things you can do to ensure proper gain structure is to make sure your sound sources are plugged into the right holes at the mixing console. There's a reason for separate mic and line inputs on your mixer: Line-level signals (such as those from effects devices, tape machines, keyboards, and drum machines) are much stronger than microphone signals. This is why manufacturers usually use different types of connectors for mic and line inputs. (Note that "tape" inputs essentially have the same gain characteristics as line inputs.) A mic input incorporates an extra gain stage to boost the microphone's feeble signal up to something more usable. Line inputs are less sensitive, so when using an XLR-to-1/4-inch adapter cable to plug a mic into a line input, you'll have to crank the gain way high just to hear the mic. (This adds more noise.) Conversely, if you plug a keyboard into a mic input, you're probably going to hear distortion because the signal from the keyboard is strong enough to overload the mic preamp. Those are examples of poor gain structure.

Audio by Numbers

Once you've ascertained that the devices are plugged into the correct jacks, it's a good idea to check their operating levels. There are variations in line level, most notably those referred to as +4 and −10. Although the boundaries have become blurred in the past 10 years, most professional audio gear operates at +4, while most semi-pro and consumer gear operates at −10. (Technically speaking for you tweak heads, it's +4 dBm and −10 dBV, but we're not gonna go there.) A general clue to operating level is the type of jacks on the rear panel: If the jacks are RCA, you can be 99.44-percent sure the gear runs at −10. If the gear has XLR connectors for the line inputs and outputs, it's almost certainly +4. If 1/4-inch jacks are used, you'll have to get out the manual and read the fine print. While you are at it, pay attention to whether the 1/4-inch jacks are TS unbalanced or TRS balanced.

It's important to understand how +4 and −10 gear reacts when interfaced together, so here's an example. You patch a consumer-style CD player into a mixer. The CD player has RCA jacks, and the mixer has 1/4-inch jacks, so you buy or make an adapter cable to connect them. But when you listen to the CD player through the mixer, the level is really low. To make it as loud as your drum machine, the faders have to be pushed way up. This is because the CD player operates at –10 and the mixer operates at +4. The mixer is expecting to receive a stronger signal level. When it doesn't, you have to crank up the faders (which generally means more noise). What's the solution? Look on the mixer for a switch that changes the operating level from +4 to −10. Doing so will make the line input more sensitive, and you won't have to open up the faders so much. (Matching tape machine output levels to tape inputs in this manner is crucial to clean mixes.) Some gear has two sets of input or output jacks for exactly this reason. If you can't adjust the operating levels of two pieces of gear to match, consider getting some sort of level-matching interface. A good example is the Matchbox from Henry Engineering. It converts −10 audio on RCA jacks to +4 audio on XLR jacks and vice versa. Similar devices are available from Ebtech, Whirlwind, and other manufacturers.

Matching operating levels is particularly important when using compressors. Let's say you have a compressor patched between your mixer and your power amp in order to prevent the power amp from being overloaded. The idea is that when the mixer starts putting out excessive level, the compressor will compress, protecting the amp and speakers. Well, for a compressor operating at −10 (semi-pro), a +4 signal from the mixer looks like excessive signal, when in fact the mixer may not be putting out much signal at all. This limits (HA!) the maximum drive to the power amp, putting a cap on the amount of volume you can get in the room. The solution in this case is to look for a −10/+4 switch on the compressor and set it to match the mixer.

The Microphone and Other Delicacies

Proper gain structure on a microphone is critical to clean sound because mics put out such weak signals. Think of a mic signal as water flowing through plumbing. Much like plumbing, audio consoles have a series of "valves" that influence the signal flow. If you require water pressure sufficient to reach the fourth floor, you have to check several valves. The most important one is the valve on the main water pipe entering the building. If the main valve is closed down, you can open up every valve feeding the various hot and cold lines throughout the building, you can open up every faucet in every bathroom and kitchen in the entire place—but as long as that main valve is closed, water will not reach the fourth floor. The mixer channel's mic trim knob is the equivalent of the water main. You must get the correct amount of level at the trim (or gain control) in order to safely deliver the mic signal to the rest of the chain. You can boost the fader up as high as you want, but if the trim is off, you'll get nothing but noise. Conversely, if you have the trim way up and the fader way down, chances for distortion are much higher.

Depending upon the mixer, there are several ways to measure the mic signal. A popular feature on many consoles is the PFL (Pre-Fader Listen) meter. Generally, pressing a button labeled PFL on the channel switches the mixer's main meter to show the level of this one channel *before* that channel's fader. In other words, it's letting you measure the water pressure right after the main valve but before the kitchen faucet. If you set the level incorrectly here, you're practically doomed to a career of distortion or noise. Adjust the trim knob while watching the meter. You can raise the trim until the meter reads 0, but remember this: Other microphone signals must make it into the audio "plumbing" during the mix, so leave a bit of headroom by PFLing the signal at roughly −7 to −5. When you start combining signals, you won't overflow the main mix pipe. Since adding EQ will likely change the PFL signal, allow a bit of room for that as well. If you have the trim all the way down and the PFL signal is still way over 0, look for a pad switch on the channel and use it; this will lower the sensitivity of the mic preamp by a fixed amount, reducing the possibility of distorting the signal (sort of like narrowing the water main).

Variations on this type of metering include "solo," as implemented on most Mackie analog consoles. The trick here is knowing that this type of solo does *not* show pre-fader level, so the fader must be set at unity, or you will not get an accurate reading of signal level at the input stage. On some consoles this spot is marked with a 0 or a small arrow. This is the spot where the fader is putting out exactly what it is receiving, neither boosting nor cutting the signal. Other consoles might have a simple two-color LED with green for signal present and red for overload. In this case, adjust the trim until the LED barely shows red and then back it off by about 10 to 15 percent. Since some consoles have more headroom than others, you'll have to experiment to see how far you can push the trim before distortion occurs.

Once the trim is set, you can bring up the channel fader to hear the signal. At least some of the channel faders should be at or near the 0 mark; if all the faders are very low or very high, something is wrong with the gain structure. Keep in mind that other "valves" affect the audio signal, such as the main mix fader(s), which should also be set at or near 0. If setting the master at 0 makes the volume in the room too loud, turn down the level controls on the power amps. If you need to bring the master fader all the way up to get adequate volume in the room, either the power amps are set too low or your system is underpowered.

When sub-mixing channels (10 channels of drums to a stereo pair of subgroup faders, for example), similar concepts apply. Think of a subgroup fader as a hot/cold mix valve. For the mix valve to operate properly, you need correct pressure of hot and cold water before mixing. Try using the kick drum channel as a reference, setting its fader to 0, and then mixing the rest of the drum channels in to taste. If the mixer has a PFL switch at the subgroup fader, use it to measure the flow right before the subgroup fader. A subgroup PFL showing in the red will probably sound distorted no matter how loud or soft the drums are in the mix.

Gain structure is equally important when using aux sends to route signal to effects such as reverb or delay. Some mixers have a PFL on the aux send. Use it. Measure the level of that snare drum send before it hits the reverb unit. Turn up the master knob for the send and watch it hit the meter on the reverb. If the reverb has an input level control, turn it up until the meter hits red and then back it down a bit. At this point it doesn't matter what the reverb sounds like; just get the level right. Then PFL the mixer's effect return to set the output level of the reverb as well as the trim on the mixer's effect return (if there is a trim). Once the levels have been set, bring up the effect return fader (or knob) to add the sound of the effect into your mix. With digital effects, correct gain structure is extremely important because if you set the input too low, you won't get the full benefit of the unit's A/D/A converters.

Once you get into good gain structure habits, you'll find that you have more system headroom, better signal-to-noise ratios, and cleaner mixes. Of course, the rules can be broken, but first it's a good idea to learn the game.

14 Aux Sends and Returns

You should now be pretty well versed on the entire signal chain, from the point on the console where the sound enters to where it exits. We still have amps and speakers to deal with, but for now let's take a little road trip into the Land of Effects. Beware: It is easy to be seduced by the pretty things native to this part of the audio world, and some of those pretty things have big, sharp, nasty teeth. So stay on the marked trail and follow your guide at all times. Let's go . . .

Hello? (hello . . . hello . . .) Is There Anybody in There? (in there . . . in there . . .)

First, let's talk a bit about where effects get inserted in the signal chain. More than 90 percent of the time, effects units will be fed a signal from an aux send on the console. Pro mixers will also offer "insert" points on each channel. So, if you have an effect or a dynamic processor that you want on only one channel, there is no need to burn an aux send. You just use the insert to put that effect on the channel where you want it and nowhere else. If you have an insert out and an insert in (known as a *dual point* insert), you use two cables and can keep everything balanced and pro. But most consoles you will use feature single-point inserts, which share a single jack for both input and output. To use this type of insert, you will need an insert cable. This is a Y-cable with a stereo 1/4-inch connector on one and with each arm of the Y terminating in a mono unbalanced 1/4-inch connector (see Figure 14.1).

The signal flows out of the channel via one part of the connector (usually the tip), which is wired to one of those mono connectors. It goes through the inserted processor and out to the other mono 1/4-inch connector, which is wired to the second conductor on the TRS (usually the ring portion). The last part of the TRS—the sleeve—is a common ground for both of the mono 1/4-inch connectors.

Back to the aux sends . . . Remember, there are two kinds of aux send—pre-fader and post-fader. Monitors use the pre-fader sends (you don't want any adjustments you make on the main fader to affect your monitor mix), while effects need to be a part of the overall sound, so they are affected by main fader moves.

There are a couple of ways to get the sound from the effects box back into the main signal path. One way is to use the returns that are dedicated to this purpose. The other

Figure 14.1 A typical 1/4-inch insert cable.

way is to run the FX output into a pair of unused channels on the console. We'll get into why you might choose one way over the other in a moment. For now, let's take a look at the kinds of effects you might use in a live setting. They fall into two main types: time-based and modulation effects.

Time-based effects are generally understood to mean reverb and delay or echo. The truth is that these are pretty much the same thing; it's just a matter of degree. A delay or echo replicates the sound of, say, your voice, bouncing off of a surface so you hear it again. Reverb replicates the numerous very short echoes that make a room sound the way it sounds. These short echoes are called *early reflections,* and the number, timbre, and volume of these reflections make up the sonic *signature* of a space. Reverb is sometimes used to make something sound bigger and more dramatic. It is best used to make a sonically dry room sound livelier. It can also be used (or overused) as an effect—such as on the vocal on a big ballad or the sound of a classic surf guitar song. Just so you have your terms straight, reverb is often referred to as *'verb* or *grease.*

Echo is a more distinct repeat of the original sound. Again, a little goes a long way, and a good sound person who is familiar with the material being played will often "feather in" a bit of delay at the end of a vocal phrase and then back off again as a new phrase begins. Echo is sometimes called *slap*. So, if you are dialing in a mix and you get asked for a short slap and a little grease, it means a quick echo and some reverb.

Getting Swishy

Modulation effects are the ones where something about the signal changes or modulates at a certain speed. This can mean anything from volume (tremolo effects usually found on guitar amps) to pitch (chorus) to time (flanging) to phase (phase shifting). All of these effects result in some kind of moving and usually "swishy" or "swirly" sound. Chorus is also used to fatten a thin sound and can be used to good effect on backing vocals.

When old guys like me were coming up, most acts were lucky to have some kind of spring-reverb effect—like those on guitar amps—built into the console or PA head. As the world has gotten increasingly digital—and the cost of digital signal processing (DSP) is now measured in pennies rather than hundreds of dollars—we have seen the birth and growth of the multi-effect unit, which can generate two or three or a dozen effects all at once from one box.

It has actually gotten even easier to add effects as the sound world becomes increasingly digital. On digital consoles, those sends and returns are virtual, not physical, and the box that houses the effects is a piece of software inserted in the signal path, called a *plug-in*. Most plug-ins are made to emulate the sound of classic processors from the past, and there is no quicker way to start a sound-guy fight than to say that a plug-in sounds exactly like the gear it is trying to emulate. Although really good plug-ins sound really good, there will always be those who *really* can hear the difference. I have seen many major touring and installed rigs, and I often see an external unit patched across a single vocal channel or across the whole mix. I know that there is likely a plug-in on the console that emulates the hardware unit in the rack, but we use what we are comfortable with and what the client needs/wants.

The major advantage of plug-ins (in addition to the fact that being nonphysical means they will not likely break down in the middle of sound check or even the show) is that you can insert multiple instances of the same plug-in on different channels with separate settings on each channel. That kind of power and flexibility can be a wonderful thing, but it can also get you in boatloads of trouble very quickly. This is where the old adage "just because you can does not mean you should" takes on real meaning. Excessive reverb, delay, chorus, or whatever *will* muddy up your sound and make it hard for the listener to do things such as understand what the lead vocalist is singing.

Clap On, Clap Off

That brings us to some almost logistical ideas that are important to getting and keeping a good overall sound.

The first is that whatever effect you are using may sound great when the band is singing, but it probably sounds a lot less great when they are trying to talk to the crowd to set up the next tune.

This brings us full circle to how the effect gets back into the console.

Any aux send intended to be used for FX will have a matching return with its own volume knob. This is a kind of master volume for the FX—the send adjustment determines how much of the signal from a given channel goes to the FX unit, while the return adjustment controls how much of the entire output of the FX device goes back to the console. The aux send on the individual channel determines how much of that particular signal goes to the FX unit.

You may opt not to use the returns and instead bring the FX output back to the board via an unused channel. The advantages of this method include the fact that if the FX output is coming back to a channel, it can easily be sent to the monitors so the performers can get a little grease without you having to buy a second FX unit. It is also an easy thing to grab a channel fader and bring it down between songs—much easier than finding the sometimes-buried return knob. Returning the wet signal through a channel also allows you to EQ the effected signal without running a separate EQ unit specifically for the FX.

Dynamics—Compressors and Gates

When you hear a really great performer or band that understands dynamics (gasp!), they pull you into the song by going from soft to loud and back to soft again. Well, a compressor does just the opposite. It squeezes, or *compresses,* the dynamic range of an audio signal. In simple terms, it does so by lowering the louder parts of a performance, allowing you to bring up the overall level and make the softer parts more closely match the loud parts. Think of it as a kind of automatic volume knob.

Most compressors have five controls—threshold, ratio, attack, release, and output or make-up gain. Threshold determines at what level the compression kicks in. Signal below the threshold passes unaffected. Ratio determines how much squeezing a signal gets. A ratio of 3:1 means that for every 3 dB of signal over the threshold, 1 dB of sound is allowed to pass. The higher the ratio, the more compression is being applied. Attack refers to how fast the compression kicks in, and release refers to how quickly it lets go. When you apply compression, you will usually lower the overall gain as you are bringing down the peaks, so make-up gain allows you to get the level back to where you started.

In general, the lower the ratio, the more natural the sound. A ratio setting of less than 3:1 is often used to add punch or to tighten a performance or a recorded track. (These are totally subjective terms, but they are impossible to really define. Your best bet is to play with a compressor and hear it for yourself.) Higher ratios are used to get the dynamics of a performance under control. Settings higher than 10:1 are called *limiting* and are most often used as system protection to keep "Loud Larry" from blowing up your speakers.

Sounds with hard transients, such as a snare drum, need a slower attack to avoid cutting off the initial crack of the drum and making it sound unnatural. Fast attack and release—combined with a ratio over about 4:1—can result in breathing or pumping. This is when the loud sound is quickly compressed and then released, and the subsequent sound passes under the threshold and does not get compressed. You can really hear the compressor working, which is something you never want to happen. As a rep from a major high-end compressor manufacturer once told me, "If you can hear a compressor working, it's set wrong or it's broken."

When to Use It

I am one of those guys who believes that as little as possible is always best, but it really depends on the act, the performance, and what you are going for sound-wise. Really good players who really listen and adjust their playing to the others onstage with them don't usually require compression. I once asked the sound guy for Tower of Power, Ace Baker, what kind of compression he used on the horn section, and he just wiggled his fingers. Those guys are so dialed in that they "compress" themselves. Ditto a lot of old-school singers who move toward and away from the mic as their performance gets softer and louder. It can be overdone, but when it's done right, we call it "good mic technique."

A common use for a compressor in a live setting is to get a soft singer or one with a very wide dynamic range up above a loud band. Another common use—one that sound guys will argue passionately about—is to put a good stereo compressor with a very light setting across the L-R outputs of the system to smooth things out a little bit. But this means you use a good (read: expensive) compressor that brings something to the party in terms of tone. Typically, we are talking about tube compressors here, and they are not cheap.

A compressor can be used to bring up a weak kick drum (but be aware that using a compressor to bring up a weak signal means you also bring up the noise floor) or to tame an out-of-control snare drum. It is almost always used on the bass, and many bass amps have a compressor built in. (Be aware that any tube amp will produce a bit of "natural" compression—a big part of the tube sound.)

Someone's at the Gate

Compressors are part of a class of devices known as *dynamics processors*. The other most common devices in this class are limiters, which we already touched on, and noise gates.

A noise gate is used to clean up a "dirty" signal and keep the noise at bay during quiet moments or, more commonly, to "turn off" a mic when it is not being used so it does not pick up signal from other sources around it and muddy up the whole sound. Drums—especially toms—are often gated so the mic is only feeding signal when the drum is actually hit. In very loud settings, it not only cleans things up, but it also lowers the probability of feedback in the drum mics. Again, be careful with the release settings to keep it from sounding unnatural. Also, a gate on a lead vocal mic will help keep the drums out of that channel when the singer is not at the mic.

There are a bunch of decent entry-level mixers—most notably from Yamaha—that have built-in one-knob compressors that allow you to dial in how much you want to squeeze the signal, and through the magic of digital signal processing, it makes the other adjustments for you. There are also some very good compressors with presets for different kinds of inputs, including the PreSonus BlueMAX and the TC Electronic C300. Those are good places to start if you are just beginning to "get" compression. Outside of the preset world, dbx is practically synonymous with compression, and some processors with other brand names actually use a dbx chip in the compressor part of the circuit.

The Secret Is Out In our *American Idol* world, there is a use for noise gates that many pro sound guys don't know about but that is common on pop acts you think might be lip-synching (and who probably are).

It works like this: Slight discrepancies between the movement of the singer's mouth and the track to which he or she is "singing" are not very noticeable to most people. But when the singer stops and the vocal track continues, you have what we call in sound-guy tech jargon "a problem." If you want a great example, go to YouTube and search for Ashlee Simpson and SNL. A classic crash and burn of a lip-synching singer...

Since that incident, some folks have figured out a way to make sure that kind of thing never happened to their clients. A noise gate is placed in the path of the recorded vocal track and is controlled by the input of the "singer's" mic. So as long as sound is going into the mic, the gate stays open, and the recorded track plays. When the singer stops, the gate closes, and the recorded sound stops. Pretty slick, huh? In other words, the singer may actually be singing, but it is not always what the audience is hearing.

---

15 Monitoring

As long as we are dealing with auxes, we're going to take a big side trip into Monitor World. Remember, I said there are two kinds of aux sends—pre- and post-fader. This refers to where in the signal path the signal is diverted into the aux. As discussed in the previous chapter, effects and processing should be affected by any adjustments to the main channel fader, so the aux is post-fader. Auxes that get sent before the fader (pre-fader) are generally used for monitoring, although they may be used to send a mix to a place other than the stage, including auxiliary rooms in venues such as churches and schools or backstage in dressing rooms.

Most consoles designed to be used as mixers for front of house will have between two and four pre-fader auxes, and in many smaller gig situations, you will find yourself running both the house sound and multiple onstage monitor mixes. (On some desks—especially some made in the UK—monitors are referred to as *foldback.*) On these kinds of gigs, if it is an analog console, you will likely deal with two or three mixes—probably one for the lead vocalist (the "star" mix), a general band mix, and maybe a special for backup singers, a second lead vocalist, or even a drum mix.

When we move into the world of digital, it gets more interesting. Even the smallest of digital desks designed for live sound will usually have many more outputs than their analog counterparts, and I know of many house engineers at venues in Las Vegas that regularly mix the house plus eight or ten monitor mixes. With the growing adoption of in-ear personal monitoring, a large number of monitor mixes has become more necessity than luxury for many artists. A "good enough" monitor mix is one thing when it is coming from a wedge at the performer's feet, but it quickly becomes not good enough when it is being pumped directly into the performer's ears.

In my experience, the job of the monitor engineer is much more difficult than the job of the FOH mixer. Not only is the monitor engineer juggling multiple mixes, he is also working much more directly with the artist, and the pressure can be intense.

The Gear

A specialized monitor desk (in the analog world) is generally known as a *matrix mixer*—every input can go to any output in varying levels. For an example, take a look at Figure 15.1, which shows a small monitor mixer that can be rack-mounted.

Figure 15.1 Crest X20RM monitor mixer. Image courtesy of Crest Audio.

Starting at the top of a channel strip, you will see controls that are familiar—the trim pots, a high-pass filter, and four channels of EQ with a sweepable mids. Next, we have levels for each of the 12 sends. On the right-hand side of the console are 12 faders that provide master output level for each of the sends. Each pair of outputs can be linked for stereo operation if you are doing in-ear monitors. When a pair of outputs is linked, they act as a single output. In other words, if you were to link all of the pairs, you would have six outputs. So if, for example, you have three vocalists in the band all using personal monitors, you could provide each of them with a stereo mix and have eight more left for mono wedge mixes.

The inputs and outputs on this console are located on the rear panel. In addition to the expected channel inputs and outputs for each of the 12 mixes, there is an output matched to each input. On a dedicated monitor board, you will find an onboard split like this. The idea is that sources come from the stage to the monitor console and then via the split to the front-of-house console.

Each output will also—most likely—have an insert point that will probably feed a 1/3-octave EQ. The biggest issue in monitors, aside from providing a mix that meets the requirements of the artist onstage, is avoiding feedback from the stage wedges. The 31-band EQ is used to lower the output of the specific frequencies causing the feedback—a process known as "ringing out" the monitors. This is one area where ear training becomes crucial. If you are going to "call" the ringing frequencies, you have to be able to identify them.

Do You Hear What I Hear?

To be anything like effective, the monitor engineer has to be able to hear as close to exactly what the performer is hearing as possible. This is where the cue wedge comes into play. The cue wedge is a stage monitor—the exact same kind as is on the stage—that is set up for the monitor engineer. If you are stuck mixing monitors from the front of house, you will still find a cue wedge very handy.

Very simply, the mix you are crafting for the performer is played for you through the cue wedge so that the two of you have a common reference point. Few performers can effectively communicate what they want in a monitor mix—but the ones who can are adamant about what they want—so being able to hear what they are hearing can make the process much quicker and simpler.

The increasing use of personal monitors has made the job of mixing monitors harder in some ways. (Side note: Personal monitors are also known as *in-ear monitors* or even just *ears,* but those two terms are trademarked by one manufacturer—Future Sonics—so although they are commonly used terms in the field, for publication we refer to them generically as *personal monitors* or *PMs.*) Although there is less likelihood of feedback than there is with stage wedges, the mix is more crucial. As someone who still performs and who uses personal monitors, I can attest to the fact that a marginal monitor mix is one thing when it is coming out of a wedge at your feet, and it is a whole other thing when it is playing inside your head.

And because a good mix and a bad mix are totally subjective in Monitor World, that means you get to do more mixes, because everyone hears differently, and no one wants to share a PM mix with someone else. Another thing: If you mix a lot of monitors, do yourself a favor and invest in at least a few sets of professional PMs. When you are mixing PMs, you will almost always—if possible—want to use the same make and model that your "star" is using so that you hear the same things he or she hears. Most touring monitor mixers I know have a half-dozen or more sets of PMs, because every time the client updates or changes, they need to do the same.

16 The Drive Rack

Before we get into this section, we need to note that we are talking about a generic set of components that together handle some processing that happens post-console and pre-power amp. For most of the history of the live sound biz, each process was handled by a separate component; put together, most of us referred to that group of components as a *drive rack*. Much changed as digital signal processing (DSP) got more powerful and cheaper, and today all of those processes are usually handled by a single component (and in some cases, that DSP is built right into the power amp).

Making it a bit confusing for some is the fact that one major audio manufacturer has a line of processors called DriveRack. So just to be clear, we will be looking at each of these processes as though each was handled by a separate piece of gear. Yes, that idea is hopelessly outdated, but I find it the best way to make sure that inexperienced sound techs really understand each process. The current crop of what we generically refer to as *speaker processors* are really powerful and, if used incorrectly, can not only make your show sound bad, but also can seriously damage the gear "downstream." In other words, you can blow up amps and destroy speaker drivers. We are going to take this approach to make sure that does not happen on your watch.

The components that process the signal after the console are largely the same as ones you would find on the console or inserted on an individual channel. However, they serve very different purposes. We are going to start with the one you won't find on the console—the crossover.

You will find crossovers as passive devices inside a "full-range" speaker cabinet, as stand-alone units, and as part of a speaker processor. All serve the same purpose—to split source audio into multiple signals based on frequency. In a full-range cabinet, the crossover really just serves the purpose of protecting the high-frequency driver. The signal enters the cabinet and is split in two (or three in the case of a three-way cabinet), and a network of old-school passive electrical components (you know, resistors, capacitors, transformers?) keep the low-frequency parts of the signal from reaching the high-frequency driver. Most passive crossovers send the full signal on to the low/mid driver. This kind of crossover is known as a *high-pass filter*. It passes frequencies above a set point (most passive crossovers are not adjustable) and blocks the rest. So what happens to that part of the signal? Remember that the audio signal is energy, and—basic physics—energy can be neither created nor destroyed, but you can change its form. In this case, the low-frequency energy is converted to heat.

The problem with this approach is simply the wasted energy—a full-range cabinet employing a passive crossover will require more amplifier power to achieve the same output as a cabinet in which the individual components are powered by separate amps. Doing this requires a crossover that comes before the amp in the signal chain. There are analog passive devices that can serve this purpose, but the chances of you coming across one are not good. In fact, the chances of running into an analog crossover at all are slim. But analog or digital, the idea is the same. A crossover basically acts as a traffic cop for audio. It stands in the middle of the street and tells the stream of cars which way to go.

Order, Order, Order!

Let's get some definitions of different kinds of crossovers out of the way. If you really want to get into the math involved, there are some great online resources, including the audio guide at rane.com (always a great reference for pretty much anything audio). For our purposes, we are going to keep it a little less math-class and a little more real-world.

A crossover is a filter, and you will see filters referred to in terms of "order"—first order, second order, and so on. Again, there is a bunch of math here, but the easiest way to think about them is in terms of how steeply they roll off frequencies above and below the crossover point. A first-order filter is limited to a 6-dB-per-octave slope, which is not adequate for most crossover purposes. Most crossovers you will find in an audio system are second- and third-order filters, which roll off at the rate of 12 dB and 18 dB per octave.

The other terms we need to understand are kinds of *passes*. A low-pass filter passes the frequencies below a set point and filters those above. A high-pass filter does the opposite and, as discussed earlier, is what you will find built into most full-range passive speaker cabinets. A band-pass filter allows frequencies between two set points to pass and filters those above and below. You can get a band-pass filter by combining a low-pass filter with a high-pass filter, and that is the best way to describe most crossovers—the combination of a high-pass with a low-pass filter. (The area where the two filters converge is often called the *sum.*) A very narrow band-pass is also known as a *band-stop*, *band-rejection*, or *notch* filter and is most often used to control feedback.

Now reverse some of those words, and you get into the guts of the filter. The pass band is the part of the signal that is allowed to—you guessed it—pass through the filter, and the stop band is the point where the filter kicks in. In theory, you would think that your high-pass and low-pass filters would be set to the same stop band, but in practice those filters usually overlap to some degree.

You will also see crossover filters referred to by the names of the people who defined them. Linkwitz-Riley, Butterworth, and Bessel are the most common. A Linkwitz-Riley filter is popular because it is the flattest in the pass band, while a Butterworth filter will give you about a 3-dB bump at the crossover point. The Bessel and Linkwitz-Riley are the most similar in that they are both very flat, but the Butterworth has the sharpest initial cutoff. The Linkwitz-Riley has moderate roll off and a flat sum. The Bessel has the widest, most gradual crossover region and a gentle dip in the summed response. The graph in Figure 16.1 gives a good picture of what the response of Butterworth and Linkwitz-Riley filters looks like when the high-pass and low-pass functions are combined.

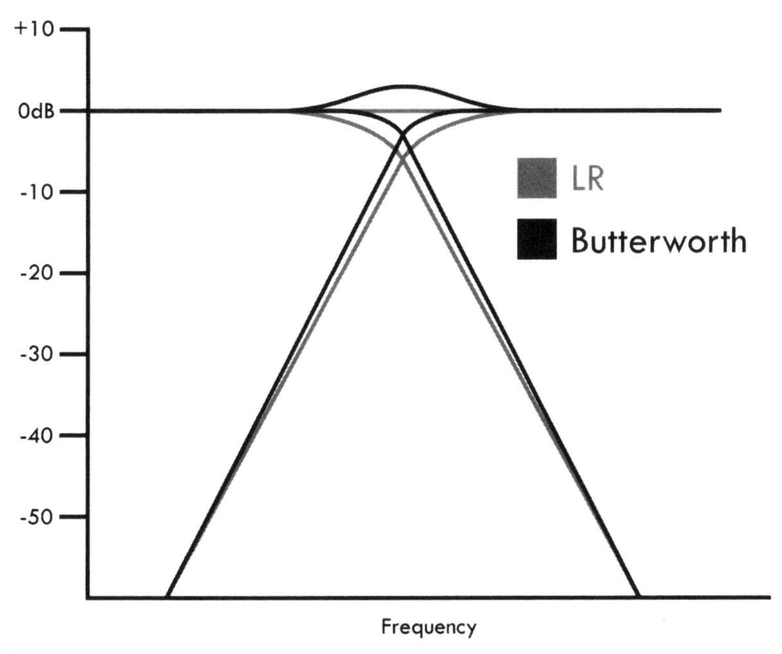

Figure 16.1 The LR line represents the Linkwitz-Riley filter, and the Butterworth line is the Butterworth. Note the area in the middle of the graph. This is called the *crossover region*, and the two filter types give you two different results. The Linkwitz-Riley filter is flat in the crossover region, while the Butterworth results in about a 3-dB "bump" in the same region. Illustration by Erin Evans.

The graph in Figure 16.1 shows the response in a two-way system. But the truth is you are unlikely to run into a lot of two-way systems. Even clubs and casual bands use three-way systems these days. As the processing gear and power amps have become less pricey and more powerful, the use of subwoofers on even small gigs has become more the norm than the exception.

System EQ

Earlier we talked about EQ at the console level, which is all about shaping tone in individual "voices" in the overall sound. In the case of system EQ, it is more about "correcting" for anomalies in the speaker cabinets or in the room itself. When I started doing sound, the system EQ was the first thing the signal hit after the console, with the crossovers coming just before the power amps. But the proliferation of cheap DSP means that we can have EQ pretty much anywhere we want it, since all of the drive components are really just different functions of DSP. It is possible to have EQ both before and after the crossover. Ditto compression and delay. As a system-wide thing, delay is used to time-align delay stacks in large venues. Now it can be used to align individual drivers in a speaker cabinet. Compression (usually set at 10:1 or higher and operating as a limiter) was used to protect amps and speakers from excessive signal input in order to keep those amps and drivers from being damaged. Now you can apply different settings to different frequency bands.

Most of the post-crossover processing falls into the area of "tuning" the speaker cabinets—EQ to make up for anomalies or deficiencies in the speakers themselves, delay to align the output of individual components—and are beyond the scope of this book. For our purposes, we have to assume you are working with a properly EQ'd and time-aligned system at least at the individual cabinet level. A bit later we'll talk about ways to align a system once it is in the room.

Getting in Tune

System EQ is sometimes called *tuning the room*. It would be more accurate to say you are tuning the system so it sounds good in the room. Depending on the gig, the amount of setup time, and just the overall attitude and vibe of the client, you may be able to run pink noise through the system (pink noise sounds like an enormous wall of static, but it contains all possible frequencies in the spectrum at equal levels) and then use a device called an *RTA*, or *real-time analyzer*, and a special measurement mic to give you a visual representation of how the room is affecting the sound. Then, you use an EQ to flatten what the mic is hearing by cutting or—if you must—boosting frequencies with an EQ. You may have to do that by just listening to a familiar piece of music through the system and adjusting using your ears alone. Regardless of the kind of gigs you do, this is a skill that you *must* develop. Yes, there *are* measurement devices that will allow you to make the system almost completely flat—as in no spikes in any part of the audio spectrum caused by the shape or construction of the room. But you won't always have those tools available, and even if you have them, you won't be able to use them on every gig. Being a competent member of the audio tribe is not about the gear. As I have often heard, it's not the car; it's the driver. The same act on the same system in the same venue can sound great or horrible in the hands of two different engineers.

Here's an example. A few years ago I spent five days on a ship called the Rock Boat. It's a very cool idea that has grown over a period of years from a couple bands taking a cruise with a couple hundred of their fans to 20-some-odd bands selling out the biggest ship in the fleet, with shows all day and night and bands ranging from indie headliners to "Hey, we just did our first CD" newbies. A ship at sea is not an ideal venue for that many shows, despite very high-end installed sound and lighting systems. Just the process of loading in and out requires the use of unusual gig gear, such as great big cranes.

The "big" room on the ship probably held 500 people, and there were at least a couple of shows there every day. Toward the end of the cruise, a national headline act did a show in that room with an opening act that was four or five rungs lower on the music-biz ladder. The opening act was mixed by one of the lead sound techs for Atlanta Sound and Lights, the production company that has been doing the Rock Boat shows since it first set sail. The headline act brought in its own engineer. The opening act sounded really good. The tech was very good. He knew his system, knew the room, and used his ears—just the way it is supposed to be.

The headline act started, and I could not believe I was listening to the same system. There was no punch in the low end and an overall dull and mushy sound that was at least 8 dB lower in volume than the opener. (This is something that virtually never happens. The politics of live music often dictate that the opening act has to run through a limiter that keeps them substantially less loud than the headliner.) The system was appropriate for the room and had sounded very good through several other shows I had checked out there over several days. But at that point it just sounded bad, and I could see that the visiting engineer and the soundco tech were having a "discussion" about the situation. What happened?

In this case, the system EQ was provided by a digital unit with very high pedigree that is used on lots of pro rigs in both the touring and installation worlds. This digital EQ basically acts like a huge graphic EQ with the potential for 100 or more user-defined "bands" and is very useful for reducing or eliminating troublesome resonances in the room. As the tech later explained, before the band even started playing, the visiting engineer grabbed more than 20 "points" on this EQ and pulled them all down by at least 6 dB—essentially cutting the amount of signal getting to the amps by about half. And then he topped it all off by screaming at the tech that the system did not "have enough gas."

We will get deeper into the roles of different players on the live audio team later in the book, but for now let's just say that, while the mix engineer will almost always be the one ultimately held responsible if there are problems with the audio, every member of the team has to be able to trust their teammates. A visiting engineer coming into a room and wanting to make changes to make their act sound their best is common. Living and working in Las Vegas, I get to know a lot of house sound teams, and I have yet to meet one that does not have a horror story about a visiting engineer wanting them to basically re-engineer the entire system for their act. Sad to say, but the great majority of the stories involve visiting engineers who don't know anything at all about system design and just want things the way they want them. But most good house crews will also be able to tell stories about the guy who came into the room for the first time and made a small suggestion that significantly improved the sound quality or system performance.

HELLO, HELLo, HELlo, HEllo, Hello, hello . . .

If, like many of us, you are coming to the audio world via the music world, you are probably used to using the terms *delay* and *echo* interchangeably. Echo is a commonly used effect and is generated by a device called a *delay*—unless you are using real vintage gear, and the hardware was still named using the word *echo*. But in a sound system, echo is a bad thing, and the delay unit is used to get rid of it so that multiple sound sources hit the listener at the same time. The difference in clarity between a properly time-aligned system and one that is not aligned is huge. Same gear, same engineer, same act, and a little delay can make things sound completely different. And this is one place where you may be able to vastly improve the sound of the system with just a few small tweaks.

Some explanation first . . . Delay in a system was first used in extensions to the system called *delay stacks* or *delay towers*, which were used to extend the reach of a sound system for very large gigs. For example, even a huge sound system is going to have a hard time throwing enough sonic energy to the far reaches of audience space at a stadium gig. So at the point where the volume drops off, a second set of speakers is set up for that extra energy needed to make it sound good in the cheap seats. The problem with this second source is it is still within the sound field of the main speakers. The electrical signal that flows to the speakers moves much faster than the sound coming from the primary speakers. What you end up with is sound from the secondary speakers followed a fraction of a second later by the sound coming from the main source. This makes everything indistinct, muddy, and dull-sounding.

The reason the secondary speakers are called *delays* is because the signal is delayed so the sound coming from the delays hits the listener at the same time as the sound from the primaries.

On smaller gigs, the issue is more about the placement of speakers, and this is especially true on club gigs. Some clubs have real systems that were installed by people who know what they are doing and spent time designing the system to be right for the room. Some are bars that book live music and pay little or nothing to bands so they can sell more beer. The system—if they even have one—may be something they bought from a local band or something they found on Craigslist and had installed by the kid up the street who usually installs car stereos. The good news in the latter scenario is that you will be making more than the band. The bad news is that you have to try to make a pile of garbage sound decent. In other cases you may be working directly with a band that carries its own PA. In any of these situations it is possible—even likely—that the system will not be aligned properly.

The importance of a properly aligned system just can't be overstated. There is a local club here in Vegas that I played at while working on this book. I was looking forward to the gig because I know the person who did the install, I know the gear involved, and I was impressed by both. I advanced the gig and was shocked at how bad the system sounded in the room—that is, until I spoke to the person they had operating and maintaining the system, and he told me that the line of subwoofers across the front of the stage had originally been behind the stage and that by moving the subs forward, he got more stage space because he put the monitor wedges on the subs. Two problems: First, I know the installer well enough to know that if the subs were placed in an unusual configuration, there was a reason for it. And, moving the monitors forward meant moving the vocal mics forward, which put them well into the coverage area of the flown mid-top boxes, turning the whole room into a feedback nightmare. When I spoke to the installer, my suspicions were confirmed. The club had run out of money, and he had to align the system physically rather than electronically because he had to do it without a processor. And he had wanted to fly the mains two feet farther forward, but doing so would have cut the coverage of the dance floor, so everything was a compromise, and changing any part of it made the whole thing fall apart.

In fact, I was impressed enough with Brian Klijanowicz, who did the original install, that I asked him to explain a couple of different approaches to time-aligning a club system. Take it away, Brian...

Time-Aligning a Club System Time alignment is a very important, yet very often overlooked aspect of system setup and tuning. A correctly time-aligned system has many benefits, including more even coverage where two sound sources overlap and a more even response across acoustical crossover points. It can give even the cheapest of systems a couple decibels more in the area where engineers tend to like them most: bass frequencies.

Two ways to quickly achieve this are using a sine wave and delaying the PA back to the kick drum. Both approaches work, and the one you choose will depend on the size of your gig as well as the time you want/have to work on it.

This is intended to be a minimalistic, quick way to time-align your system, so only a few pieces of gear will be required. Using a sine wave will only need a sine wave generator and the capability to delay an output signal and to reverse the phase of it. To delay the PA back to the kick drum, all you need is a channel of delay and your "golden ears."

We will assume that the sound system we are working with is a standard front-loaded "stack" configuration. This means there is a subwoofer (producing sub-frequencies from roughly 100 Hz and down) ground-stacked on the floor with a top box (producing roughly 100 Hz and up) directly on top of it. We will also assume that both subs and tops are currently producing the same polarity and facing the same direction.

Sine Wave

Usually, an engineer would not want to use polarity to cancel out a signal. But that is the whole concept behind this technique.

First, it is important to find the acoustical crossover frequency between the subs and tops. A measurement device is the most accurate way to do this. But if you don't have one, there is an easy way to rough it in. Assuming you don't have a measurement device, flip through the pages of your crossover to see what the crossover frequency is and use that. (For this article, it will be 100 Hz.)

Let's get back to the concept. We will take two similar sound sources, sub and top, that are playing a 100-Hz sine wave, and we'll flip one out of phase. The sound sources will overlap and start to fade off from one another at the crossover point, but more importantly, they will still reproduce 100 Hz, because that is the beginning of the crossover filter on each source.

By physically looking at the speaker cabinets and knowing where the drivers are in the boxes themselves, you will be able to determine which signal to add delay to. Typically in the club world, top boxes are placed a little behind the front of the sub. That way, if the top box falls, it will fall back and not onto the drunk audience. So, assuming this is the case, delay should be added to the sub. While the 100 Hz is playing through the sub and top, start increasing the delay to the subwoofer signal. Eventually, the two signals will start to cancel out, and the total SPL will reduce substantially. Find the point at which most cancellation occurs and leave it at that. Change the signal that's out of phase back into phase, and you should have summation at the crossover point.

Some crossovers have an on and off function for the delay. A good way to check whether summation is occurring is to flip the delay on and off to hear the difference.

This whole process can sometimes be done with music, preferably with driving, kick drum–heavy music. However, it can be hard to distinguish what is cancelling and summing in the crossover region with full-range music playing. A quick fix on a digital console would be to throw a low-pass filter on the music channel.

Back to the Kick

Delaying the PA back to the kick drum is very simple. It works very well in the small "hole-in-the-wall" biker bar, but once you get into a bigger venue, it can become a bit harder to hear the difference. Smaller venues that have lots of reflective surfaces and 90-degree walls create standing waves. This makes it almost pointless to spend the time to get out a measurement device—especially when the club owner wants you to be done setting up before the first road case is even in the building.

Since this technique works better in small venues, we will assume your PA is in a small venue and set up in front of a band. Even though it's still surprising to club owners and patrons, we all know how loud a drum set can be in a small room. That is why typically, most engineers won't even turn up all the drum mics, or they just won't mike the whole kit.

So the drums are roughly 5 feet behind the PA. Think of the kick drum as another speaker. If you are sitting in the audience, would you want to hear the main PA with another set of PA 5 feet behind it? Probably not. So if you are sitting in the audience listening to a band, you probably would not want to hear the kick drum once through the PA and then again 5 feet later.

Most drummers do relatively the same sound-check routine. They will hit each drum individually with quarter notes at a medium pace until you ask them to switch to the next drum. When you get to the kick drum, get a rough sound in on the channel strip and start tweaking the delay back. Add a little bit of delay at a time. You will start to hear a change in tone and depth. Once you get to the sweet spot and are happy with the sound, that's it!

Take It to the Limit

The final step is to put some kind of "emergency brake" on the system—especially if it is going to be driven by visiting engineers—in order to protect your amps and speakers. Basically, this means using a compressor set to 10:1 or higher or a "real" brick-wall limiter so that the amount of signal reaching the amps is below the level at which the amp clips, which can damage both the amp and the speaker.

Now, let's go back to the beginning here and remember that we used to have to deploy a separate piece of gear for each of these processes. But the current crop of digital speaker processors put all of these functions into one box. A typical processor will have two or four inputs and six to eight outputs. Each input signal can be split, EQ'd, delayed, and limited before moving on to the amps. Also, more and more of these functions are moving directly into the power amp, and most major amp manufacturers sell at least one model that has DSP processing in the amp. This leads to a conundrum when it comes to buying gear. You only need one set of processors. What if you have outboard processors and it is time to buy amps, and the amp already has DSP? Do you sell off your processors and go with what is in the amp? Or do you keep the processors and have the DSP in the amp as backup? To make things even more complicated, some speaker companies insist you use specific processors that are preprogrammed with the cabinet tunings for their speakers, and they will not release the data for those settings so they can be transferred to the processors you may already own. (But note that the limiter settings are usually left open.)

Just an opinion here, but this kind of attempt to keep data proprietary feels an awful lot like the battle that the record industry fought and lost a decade ago. Rest assured that companies who make processors will figure out the proprietary settings and pass them out freely to people and companies who buy their "non-approved" processors to use with whatever speakers they want to. There are only two ways to keep this from happening. The first is to only sell your products as complete systems—amps, speakers, and processors as a single unit. Several companies have taken that approach, but when the pressure is on to make sales, those written-in-stone policies of selling only complete systems tend to get a bit softer, and I know of only one company that consistently refuses—even on large systems—to sell individual components without the other parts of the system (except for replacing something that has worn out or failed). The other way to keep this from happening is to build all of the processing and amplification into the speaker cabinet itself. And powered or active speaker systems are becoming increasingly common. Which leads us to our next chapter . . .

17 Active Speakers

Before we move to the end of the "traditional" signal chain, we need to take a little side trip to a land where speakers, amps, and processing all live in the same box. A magical place called the Land of Active Speakers . . .

If you have been paying attention, you will have noticed a thread running through much of our content so far that involves a consistent move toward doing more with less. Well, kind of. The power available has grown exponentially. When Bill Hanley did the sound at Woodstock for somewhere north of half a million people, he had less total power available than your average bar band has in their music-store-grade PA. But in general, the trend has been more power—be it power or processing or whatever—in fewer boxes.

It's the Law

As we looked at in the previous chapter, most of that is in pretty direct relation to Moore's Law. (Gordon Moore is an engineer who, many years ago, while working with chipmaker Intel, posited that the computational power of a given device would double while the cost to produce it fell by half every 18 months. And that has been pretty accurate for some three decades.) The ever-increasing speed at which technology advances has made it possible in a pretty short period of time to do with one box what used to take four boxes to achieve.

This is both positive and negative. On the positive side, we get a ton of new toys and are at the point where even regional acts in marginal venues can sound better than top headliners did just 10 years ago. The negative side to all of this is that, as a former touring sound guy who is now an exec with a major audio manufacturer told me, "We have adopted the product cycles of the computer industry." He is right, and the implications are huge. Two years before writing this book, I did a buyers' guide for *Front of House* magazine on digital consoles with a price tag that made it possible for the average regional sound company to consider owning one. A few months before I wrote this, we did another buyers' guide, but this time we focused on the big, state-of-the art units. In both cases we set the same price point—$75K. On the first go 'round, we had companies screaming that the price was unrealistically low. Less than two years later, we could find only a handful of consoles above the price that had been "unrealistically low" not long before. At the time of the first guide, a digital console cost as much as a house in many places in the U.S. By the time the second guide rolled around, I knew of at least a handful of touring sound guys who owned their own consoles and rented them back to the band they were touring with because the price/performance ratio had fallen to the point where buying a pro digital console was more like an investment in a good car than in a house.

So what is the negative side? If you end up starting your own sound company, it is very hard to know what to invest in when it comes to new gear. By the time you are working in the field, it is very possible that most regional companies will have given up on trying to stay current, opting instead to keep a few workhorse consoles in their inventory and rent when they get an act or a venue that insists on something else. It is already at the point where some system installers are telling performance venues that deal with touring acts to buy a really good analog console instead of the latest in digital technology. The theory is that the lower-tech board will be perfectly adequate and acceptable for at least half of the acts that come through and that when a larger headliner is booked, there's a good chance that the digital console they chose for the install would not be on the rider (the list of acceptable gear for any gig), and they would end up renting one from elsewhere anyway.

When you look at it objectively, it makes a lot of sense. But it also means that console manufacturers are all competing for pieces of a smaller pie, which impacts profit margins, which impacts R&D budgets, which impacts how many new tools we get and how often we get them.

Back on Track

Sorry, I got a little off track there . . . Important stuff, but what we started out with was trying to make the point that putting the speaker, processing, and power all in one box was less a function of implementing new technology than it was of making the currently available technology easier to set up and run properly. Designing a sound system that works to its full potential without damaging individual components means having a firm grasp of the physics of sound, including the way sound waves propagate from a source, how sound sources work together or interfere with one another, the behavior of electro-mechanical devices such as speakers, and the power needs and potentials of the devices feeding them. This is where classes on electrical engineering and acoustics come into play. The concepts and knowledge those classes impart are well beyond the scope of this book. (By the way, that is about the fourth time I have typed that phrase, and we are not yet two-thirds finished. Are you getting the point that doing live audio right means plenty of learning of both the book and real-world varieties?)

Matching the right amp with a given speaker requires more knowledge than even people who have been around live sound for a long time may have a firm grasp on. Al Siniscal founded the pioneering live audio company A-1 Audio and is one of the two people most commonly cited as being the father of powered speakers for live event audio—the other being John Meyer, who deployed and integrated a system of amps, processing, and speakers for a 1971 tour. The disagreement is largely semantics, because Meyer, then working with McCune Audio in the San Francisco Bay Area, designed and integrated the system, but the speakers and amps/processing were in separate boxes. Siniscal was mounting crossovers and BGW power amps right into the back of his speaker cabinets and did his first international tour with those boxes in '74. Both were trying to overcome the same set of problems, mostly rooted in ignorance of the science of sound. In a 2003 feature in *Mix* magazine, Siniscal recalled a symposium sponsored by a major gear manufacturer at the 1992 Audio Engineering Society convention, where engineers and dealers were asked how they powered a set of then-popular portable PA speakers. There, he is quoted as saying, "For the five guys that can tune it, there's another 95 that will screw it up. By the time we got around the room, there were all sorts of things people used to power the speakers," he says. "These would include cheap, high-frequency amplifiers without overhead capabilities or dynamic range. They used all kinds of off-brand crossovers and processors. They wouldn't necessarily use the big enough amplifier for the low frequencies, but they were out there competing. The preponderance of people didn't power the speakers correctly, nor use the processors that the factory recommended. These were music-store items, unbalanced, typically marginal units."

From the time of their introduction in the early '70s until the mid '90s, powered speakers were largely confined to studio monitors (in fact, the first actual powered loudspeaker Meyer made was meant as a studio monitor until the Grateful Dead used them on tour) and Meyer Sound, plus guys like Siniscal who were making their own boxes. In 1995, JBL introduced the first of what has become a ubiquitous class of gear: Active systems mounted in a molded plastic cabinet that could be mounted on a tripod stand. You will often see them referred to as *speakers on a stick*. It is usually a derisive term, but I know of very few sound companies that don't keep some in their inventory for small gigs, extra monitors, or a million other uses. The market continued to grow, but it really exploded with the introduction of the SRM450 by Mackie. Today, it is the exception rather than the rule for most companies not to offer powered versions of their speakers. And there is even an outboard speaker processor of the type we talked about in the previous chapter, intended to be used specifically with powered speakers.

Pros and Cons

The advantages of powered speakers are many, including taking less truck space to move because there are no amp racks to carry, as well as being secure in the knowledge that amps are properly matched to speakers and that the processing (delay, crossover, EQ—all of the things we discussed in the last chapter) is hardwired in so that the box sounds as good as it is able to right out of the box. Another biggie is that the cable runs from the amp to the speaker are short enough to be negligible in terms of impedance and the power loss you can get when running long lines between your amp and speakers.

But there are disadvantages as well. First, you have to get AC to each individual box instead of to a single amp rack, and second, if you do lose an amp, you have lost that box—probably for the duration of the show. Especially with flown systems, getting access once the show is in progress is often impossible, and even if you *could* access the box, you would have to either replace the amp module or have an external amp and processor with all of the tunings already dialed in. Bottom line: It ain't gonna happen.

Finally, while the cable runs between the amp and speaker are very short, the run to the speaker will probably be as long as the initial run to the amp (and longer if amps are located at the front of house—an undesirable but not unusual situation). While long runs between the speaker and amp can have their own issues as outlined earlier, at least the signal flowing through that line is strong enough that it is far less likely to pick up extraneous noise. On the other hand, that long line-level run to a powered speaker is a scenario that almost begs for interference coming from anything ranging from overhead power lines to a radio station or any other wireless transmission.

For this reason, it is imperative that you run balanced lines between your console or stage box and any powered speakers. I once did a small church gig that started outside in the parking area, which just happened to back right up to a small radio station, before moving inside for the second half of the event. I was worried enough about interference that I used no wireless mics at all. I thought I had it covered, but when we fired up the system, the radio station was coming through loud and clear. The problem ended up being simple and was easily fixed, but... not to leave you hanging or anything, but I am not going to get into it right now. We'll save it for later in the book.

To close out this subject, I am going to tell you not to do something the whole industry does and that I have in fact done throughout this chapter. Try not to confuse or interchange the terms *powered* and *active*. A powered speaker has an amp in it and maybe a passive crossover. If you want the advantages of a speaker processor that we talked about in the previous chapter, then you still need to run said processor between your source (usually the console) and the speakers. On the other hand, an "active" speaker at least implies that the processing is handled in the box. A real "active" box will basically be pre-tuned, and many are made to be matched as specific sets of subs and top boxes.

A real-life example: About five years before writing this, I was doing a gig with an active system that I had used for a few years. The crossover was built into the sub, so I just ran signal to the sub and then looped it to the top box. This is a very typical setup for smaller gigs, and you will come across it often, especially in small clubs. A friend had been touting the services of a company that used a very high-end and precise audio analyzer to measure a dozen different audio parameters and then make suggestions on how to better tune the cabinets. This was a midsized sound company with an inventory that included both powered and unpowered boxes and that used external processors on all of them. The owner told me that the tweaks recommended after the analysis made his speakers sound better and gave him a few dB of extra gain.

So I had them come out to the gig to analyze this active system and see whether it would benefit from an external processor. We set up the system using the external processor as a crossover only set at an arbitrary 100 Hz. He ran his tests, and the system was a mess. Drivers were out of phase, the response was so far from flat that the meter looked something like a mogul run on a ski slope, and in order to be aligned without a delay, the subwoofer would have had to be set 9 feet behind the mid-high top box.

It was hard to believe that the system was so out of whack, so we did the analysis again, this time with the system set up as intended with the sub feeding the top boxes. The result? While still not totally flat, the response smoothed out tremendously across the spectrum, the out-of-phase drivers reversed, and the sub and top box were suddenly aligned.

Could we have gotten the same or better results from a good external processor? Probably. In fact, almost certainly. But could we make it enough better to justify the expense of the external processor? Probably not.

So what should you take away from this chapter? The fact that having the amp inside the box has its advantages, especially for those who still have a lot to learn about sound systems. But there are disadvantages as well, and those disadvantages make some pro sound guys unwilling to give up the control they have with separate components. Also, the fact that active and powered are not the same thing... High-end powered systems are generally deployed with the assumption that they will be used with a processor. Finally, although it is surely possible to improve on the response of an active system through bypassing the onboard processing and tweaking it with a processor (assuming you have the knowledge and the means to take the measurements in the first place), doing so is probably not a cost-effective option.

18 You Gotta Have Power...

We are getting near the end of the signal chain—approaching the transducer that will convert the electrical energy (or data, in the case of digital mixing) back into the acoustical energy that originated at the mic. But we're going to do it on a much greater scale (as in a lot louder). And making it loud takes power—a bunch of it. Yep, you guessed it—it's time to talk about power amps.

Let's just make sure we have our terms straight first. A power amp is not the same as, say, a guitar amp. It has no preamp or tone-shaping function. (Well, with the current move toward including system DSP in the amp, that is not really true, but I am trying to keep it simple here, so I am confining the discussion to traditional power amps.) A power amp has one job, which is to make the incoming line-level signal powerful enough to drive the loudspeakers and "make" sound. You will see units out there with digital readouts and all kinds of other controls and indicators, but the truth is that a real power amp needs just four things accessible to the user—inputs, outputs, a gain control, and an on/off switch.

Ins and Outs

Inputs and output connections on power amps come in a bunch of flavors, and once upon a time you could make generally accurate estimations of how "pro" an amp was by the kind of connectors it used. For example, if an amp had RCA inputs, it was consumer grade or at best something designed for the home or installed market. If it was out on a live show, you probably had a problem. But today, one digital format uses RCA-type connecters as an option, so that benchmark is out the window.

Today you will find a couple of different analog input types and several possible digital connections. On the analog side, what remains true is that pro gear takes balanced input signals. It can be on a barrier strip, a 1/4-inch tip-ring-sleeve connection, an XLR, or a combo jack that will take either one. RCA analog or unbalanced 1/4-inch tip-sleeve connectors denote non-pro gear.

On the digital side, it can be an XLR, an RCA, or an RJ-45 (Ethernet) connector, depending on the amp and manufacturer and the kind of digital signal the amp can take. As I noted back in Chapter 4 on cables and connectors, you really do need different cables for the digital connections. In the case of an RCA or phono connection, those are used for the S/PDIF digital format and require the use of a higher-quality 75-ohm cable. (Typical consumer RCA cables are more like 35 to 50 ohms.) The higher-rated cable simply passes more data, and using cheaper cable can really degrade the quality of a digital connection.

On the XLR side, it is a similar thing. The format is going to be AES/EBU (much more common in pro use—especially in live sound—than the S/PDIF format), and a high-quality 110-ohm cable is the spec for this kind of connection. At the time of this writing, on the RJ-45 tip we were just starting to see really pro-quality Cat-5 or Cat-6 cable making its way to market. While substantially more expensive than the cheap networking cable you get at RadioShack, it is worth the investment in a portable live sound application. If you are installing a system and the cable is not to be moved, then any decent networking cable should do the trick, but keep in mind that these cables and connectors were never made with the intention that they be used for anything other than a pretty permanent connection. Your standard RJ-45 connector is designed to withstand being plugged in and unplugged maybe 50 times, which is why everyone I know who has gear with an RJ-45 connection either invests in better cable or carries multiple backup cables to every gig, because those standard networking cables are *going* to fail... and probably sooner rather than later. On the connector side, you are always best off using gear that will take a Neutrik dataCON, which I discussed and showed a picture of back in Chapter 4.

Outputs can take several forms. Even though few speaker systems use them anymore, many amps still include banana-plug or binding-post outputs. These are made to take a bare wire or simple plug that uses a separate plug for the hot and ground connections.

A straight 1/4-inch jack may be available, but again, this tends to be found on less-than-pro gear. The most common output connector is the Neutrik speakON (see Figure 18.1), variations of which are made by other manufacturers and called *twist-lock*.

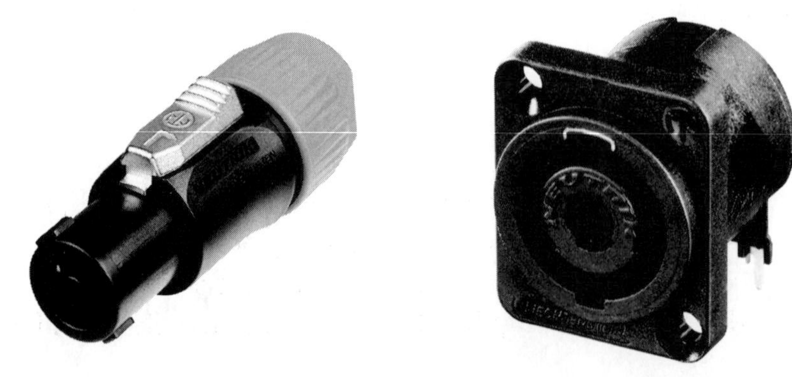

Figure 18.1 Male and female Neutrik speakON connectors. Images courtesy of Neutrik.

speakONs come in flavors that include a straight two-pole connection, like a 1/4-inch connector, to NL4 and NL8 connectors with four- and eight-pole designs that allow for bi-amp or tri-amp connections on a single cable. (Actually, with eight conductors, an NL8 could handle a four-way system on one cable.)

What's Cooking Inside?

We were talking about amps and got kind of sidetracked into cables and connections again. (But keep in mind that more than 90 percent of the time you are looking for the point of failure in any audio system, your search will eventually lead to a cable or connector.) Power-amp technology has made huge leaps in just a few years. When Bill Hanley basically invented this business in the '60s, one of his biggest competitive advantages was a refrigerated truck that held all of the Macintosh tube amplifiers and kept them cool enough to power a large gig. The Woodstock festival was done with maybe 1,000 watts of total power. Today, that is one channel of a smaller pro power amp.

Amp design very quickly gets into electrical theory and formulas that are well beyond the scope of this book. But we should at least take a look at the various amp classes before getting into some of the more practical knowledge you will need for doing this stuff on a day-to-day basis.

Class A

Those tube amps I referred to a moment ago were a Class A design. Class A amps sound really good and are prized by audiophiles to this day. But they are inherently inefficient. That is, the ratio of output power (what drives the speakers) to input power (what you plug the amp into so it will turn on) is low, and much of the available energy is wasted and dissipated in the form of heat. Remember basic physics: Energy can be neither created nor destroyed, but you can change its form. In this case the electrical energy that powers the amp is sort of transferred, for lack of a better term (and amp designers reading this are shaking their heads in disgust right now because that is not really accurate, but it is easier to understand without the math you need to really comprehend the inner workings of an amplifier), to the input signal, making it many times more powerful at the output than it was at the input. So Class A sounds good but does not make efficient use of power.

Class B

Class B is much more efficient than Class A, but the design introduces changes to the signal (in other words, intermodulation distortion) that make it sound less than great. Class B amps were most often used in applications where the quality of the sound was not a primary concern—things like portable tape machines and AM radios popular in the pre-Walkman era. They are also found in things like bullhorns, walkie-talkies, and emergency vehicle sirens.

Class AB

These were the first amps really designed with the high power and good sound needed for live audio in mind. These are big, heavy beasts that do a very good job of amplifying sound in a pleasing manner, with much greater efficiency than a Class A design. But they require a large output transformer as part of the design, which makes them less than fun to carry around. As an example, a Crown Macro-Tech 2402 will put out about 500 watts into a "standard" 8-ohm load and weighs in at about 50 pounds. Given that all but the smallest of backyard party gigs will require at least two amps, you now have an amp rack that weighs well over 100 pounds (including the weight of the rack itself). There are significant weight and size considerations—outside of the strain on your back, big, heavy amps require you to have more truck space and burn more diesel to get from gig to gig. A significant lowering of size and weight would make the financial equation for touring sound a lot easier to handle. It is not uncommon for a large touring show to carry somewhere in the neighborhood of 100 amps. That is 2 1/2 tons of amps distributed among some 16 rack cases. Add those cases and all of the internal cabling, and you are probably north of three tons. Reducing the space and weight by half would lead to very significant cost savings for touring shows. Which leads us to . . .

Class D

Though first designed in the 1950s, the components needed to build a Class D were too expensive to be used in mass production until around the end of the '90s. Class D amps operate on very different principles than more traditional power amps and are often referred to with terms such as *pulse-width modulation, switch mode,* and *switching power supply*. They may sometimes even be referred to as "digital" power amps, even though the generation of power is an inherently analog process. The bottom line is that they offer at least the efficiency of typical Class AB designs in a much smaller and lighter package, and most of the touring amps you see now are Class D designs. Crown uses a variation on the Class D pulse-width modulation design, which they have designated as Class I. Their popular I-Tech series uses this topology. Another advantage of the smaller and lighter vibe is that flown PAs can have the amp racks flown right next to the speakers, keeping cable runs short. And when it comes to cable runs between the speaker and the amp, the shorter the better.

In a perfect world, we would have perfect amplifiers that did nothing but increase the energy of a signal without affecting the character of that signal in any way. But this is not a perfect world, and there is no such thing as a perfect amp. All amps affect the input signal in some way that introduces artifacts that are generally lumped under the common designation of distortion.

There are two main kinds of distortion to be concerned with. The first is harmonic distortion, which basically introduces additional frequencies that are evenly related to the original frequency. Small amounts of harmonic distortion can actually be pleasing to the ear and are generally what is being touted when people speak of a piece of gear sounding "warm." The second type of distortion is intermodulation distortion, which is similar to harmonic distortion except that the additional frequencies are not related to the original in a way that any of us want to hear.

There is a third kind of distortion that comes into play with any amplifier: clipping. If you were to look at an audio frequency as seen on an oscilloscope, you would see the tops and bottoms of the smoothly shaped waveform get lopped off, producing a flat spot. This happens when the amp is pushed past the power it is able to supply and it runs out of gas at the top and bottom of the signal, clipping those parts off of the waveform. The overdrive and distortion used widely by guitarists is a kind of clipping and not something you want coming out of your PA. In addition to sounding bad, clipping makes an amp run hotter, which can cause it to overheat and shut down if it has protection circuitry and just burn out if it does not. Driving the input signal of the amp into clipping can cause the same problem in the speakers to which it is attached. That's right. You can blow speakers not only by feeding them more power or power at frequencies they were not made to handle, but also by feeding them with an amp driven into clipping. In fact, it is more common to blow speakers from using too little power than from using too much. Think about that when you are thinking about skimping on power amps.

One other function of an amp that can be affected by misuse is called the *damping factor*. Again, too much math, but suffice it to say that changes in the damping factor can result in fewer low frequencies being produced, making the audio sound thin through the PA no matter how dense the original sound is. One big thing that can adversely affect the damping factor is using thin or low-quality cable between the amp and speakers. When it comes to speaker cabling, use the best and beefiest cable you can afford.

What a Load . . .

Power ratings on amps can be very misleading and are often more marketing than science. When you look at the power rating on an amp, there are a few things to take into account and make sure you are comparing apples to apples.

To begin with, there are several ways to express output power. Until the marketing departments and accounting suits took over the audio business and it became all about making your gear *appear* to be better than the other guy's by using whatever irrelevant or inappropriate number it took to get there, every amp used the same spec to express output power, called *RMS* or *root mean squared*. RMS is basically a kind of average power output over time, and although none of us knew the math behind the number, we didn't need to because everyone was using the same objective basis for their ratings.

But now you need to be careful, because even ratings with the same name can be arrived at by different means. The other most common ratings are peak power and continuous or program power, which sound the same but are usually arrived at by two different routes. Peak power is the amount of power the amp can put out in short bursts. This number is obviously much higher than the RMS. Both continuous and program are also generally higher than RMS as well. Continuous usually refers to the output power an amp can put out continuously given a specific input frequency and power. But if two amps have the same continuous rating but were arrived at using different frequencies, then both numbers are meaningless. Program can be similarly meaningless because although it should be more "real world" than any other rating, no one is specifying what kind of program material is being used. Is it a solo acoustic guitar or bass-pounding hip-hop? Because of the inconsistency in the basis for power ratings between manufacturers and even among models or series from the same manufacturer, it becomes even more important to use your ears and input from hopefully objective third parties in making your decisions.

Another thing that will have a huge impact on power ratings is the ohm load. A single speaker cabinet will usually present a load of 8 or 4 ohms, and the output power of the amp goes up as the load goes down. Given that, many manufacturers give their top power rating based on a 2-ohm load. A couple potential problems here: First, while many current power amps are designed with a 2-ohm load in mind and can handle it, many (if not most) amps made before about 2005 may be able to run on a 2-ohm load, but likely only for a limited amount of time. As that output power goes up, so do both the current draw (the amount of current needed to drive the amp to that top rating) and the amount of energy dissipated as heat. I once did a gig and had an old amp just die during sound check, and in desperation I rewired the system, which meant the other amp was running at less than 4 ohms. The result? I ended the night with *two* blown amps instead of one.

And that 2-ohm rating is meaningless if you can't get enough power out of the wall to drive it that hard. As a good friend of mine at a major manufacturer once said, some of those 2-ohm ratings are only good if you are powering the amp with "solid copper bar connected directly to the Hoover Dam." Again, something to keep in mind when shopping...

Oh, and another marketing ploy: Is that rating power per channel or is it with the amp in bridged mode? This is important because in bridged mode, the two channels combine into a single output at close to double the power of either channel on its own.

You Can't Do That with an Amplifier

If things continue the way they have been going lately, then by the time you read this, you may be hard-pressed to find an amp that does not have onboard DSP to perform most (if not all) of the functions of an outboard speaker controller. Over the course of just a few years, DSP included in the amp has gone from the province of only the most expensive, high-end pro amps to being almost ubiquitous to the point where even amps designed directly for the musician market include some kind of limited system control functions.

Okay, let's wrap this up with a few rules of thumb.

1. Class D (and Class I) amps are generally more expensive, but they are smaller and lighter than Class AB amps, which could have a long-term effect on the bottom line.
2. Use the shortest, heaviest cables you can between the amp and speakers.
3. Use an amp appropriately matched to the speaker. An overpowered amp can blow the speaker through sheer power, and an underpowered amp can damage the speaker if it is driven into clipping in search of more volume.
4. A lower ohm load can mean more power, but don't go lower than four unless you are sure the amp is designed for it.

19 Loudspeakers

It's been a long journey, but we have finally arrived at the end of the signal chain. Remember that at the beginning we converted acoustic energy via a microphone; sent it on through the console where it was tweaked via EQ, dynamic processing, and maybe some additional effects; mixed it together; and sent it along to be amplified. Now we are going to convert that energy back into acoustic energy via another transducer—the loudspeaker.

Speakers can get very complex, and this is the area where acoustics and the physics of sound come into play more than any other. Getting really deep into that is outside the scope of this book, but I will give you the basics. This will end up being the largest part of this book, and I will approach it from the inside working out. First, we will examine the actual loudspeaker transducer units or *drivers*, and then we'll move on to how the design and construction of the actual box affects the sound, how you use the device, and finally, some common ways speakers are actually deployed in typical live sound situations.

The Drivers

Any kind of loudspeaker is an electromechanical device. In other words, it has moving parts. And anything with moving parts is prone—to some degree or another—to wearing out and breaking down. This can actually be an opportunity and job security for newbie sound techs. Learning how to troubleshoot and repair/re-cone speakers will put you in much greater demand than someone whose only skill is mixing. Remember, there are seven days in a week, and for the majority of sound companies, only two or three of those are actual gig days. The rest of the time is all about prep and making sure the rig is working right. Taking a class on speaker repair may be one of the best moves you can make when it comes to staying employed even when things are slow.

We are going to limit our discussion of driver types to the most common ones you will see—cone-based speakers and compression drivers. Other driver types, including ribbons and piezo, are out there (and ribbons are becoming increasingly popular, especially in line-array deployments), but the vast majority of boxes you will actually use will have these two component types inside.

We'll start with a typical woofer. The diagram in Figure 19.1 shows a speaker made by JBL.

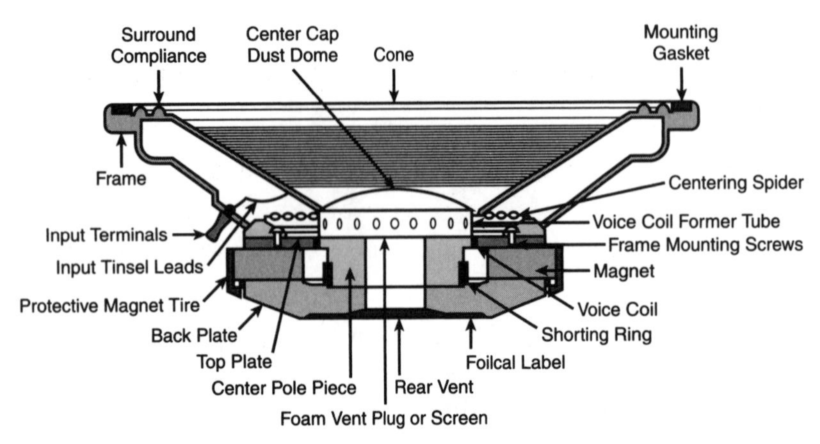

Figure 19.1 Cutaway drawing of a typical JBL low-frequency transducer. Diagram provided courtesy of JBL/Harman.

At its most basic, a cone speaker consists of a frame or basket that houses and supports the cone, a surround (flexible surface—rubber or polypropylene or a similar material) that connects the cone to the frame, a permanent magnet and metal parts forming a magnetic gap, and a voice coil. The voice coil creates an electromagnet. The incoming signal causes the electromagnetic voice coil to change polarity hundreds of times every second. Suspended in a magnetic field formed by the permanent magnet, the coil moves toward and away from the magnetic gap as the polarity changes. (Assuming a positive charge on the permanent magnet, when the coil is negative it is attracted to the magnetic gap and moves toward it, and when the charge of the coil is positive it is repelled and moves away from the magnetic gap.) The voice coil is attached to the cone and causes the cone to move in sympathy with it. The resulting motion of the cone moves the surrounding air, which reaches the ear and is perceived as sound.

Those are the basics. Let's take a look at a real speaker. Figure 19.2 shows pictures of actual speakers, and we will be using the diagram in Figure 19.1 as a reference point. Starting at the top-left and working in a counterclockwise direction, let's look at the individual parts of the speaker.

- **The frame.** This is the metal housing that supports, protects, and houses the actual cone.
- **Input terminals.** These two input points (one positive and one negative) are where the amplified signal enters the speaker mechanism. The input tinsel leads carry the signal from the inputs to the voice coil.
- **Voice coil.** This creates the electromagnetic field that interacts with the permanent field from the magnet in the gap.

- **Magnet, protective tire, back plate, and top plate.** The magnet is the source of the permanent magnetic field, the tire surrounds and protects the magnet, and the back plate and top plate form the magnetic path to the gap in which the voice coil is suspended.
- **Center pole piece.** This is the piece that the voice coil is "wrapped" around and completes the magnetic gap circuit path.
- **Vent.** This allows air to escape the structure as the coil moves toward the magnet. This air movement is all about keeping the voice coil as cool as possible. Some speakers also use a metallic fluid called *ferrofluid* in this space to absorb heat and keep the voice coil cool.
- **Shorting ring.** This reduces distortion induced by the action of the magnetic fields.
- **Former.** This keeps the voice coil in shape (not as in going to the gym in shape—as in round in shape).
- **Spider.** This keeps the coil centered.

Figure 19.2 Back, side, and front views of a cone driver, courtesy of Eminence.

Note that there is a structure both inside and outside the voice coil that forms the magnetic gap. The amount of clearance between these structures and the coil is tiny, and the coil can expand with the tremendous amount of heat generated by the power through the coil. As a result the tiniest amount of misalignment can result in scraping and damage as the coil moves. Yes, this means that a poorly designed, badly built, or improperly used speaker *can* catch fire. Something to keep in mind...

Figure 19.3 presents a diagram of a voice coil off center and tilted within the air gap.

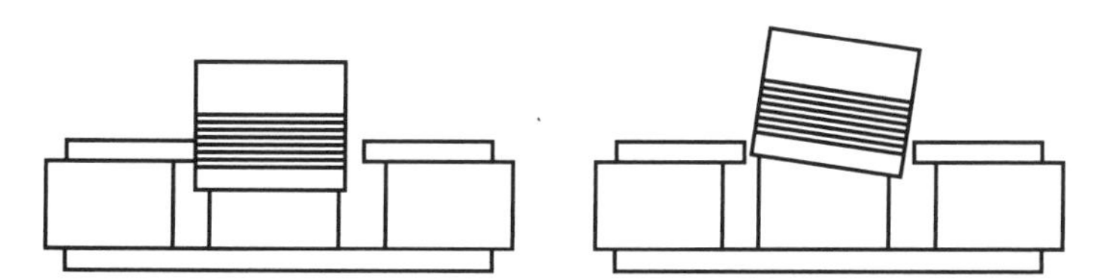

Figure 19.3 Off-center and tilted coils. Diagram courtesy of Eminence Loudspeakers.

The long wavelengths of low frequencies mean that the cone has to really move in order to re-create the bass tones. A high-frequency driver reproduces sounds with much shorter wavelengths, which therefore need to move much shorter distances. A compression driver works on the same principle, but instead of a large cone that is inverted toward the magnet, the surface is a dome-shaped diaphragm that protrudes away from the magnet.

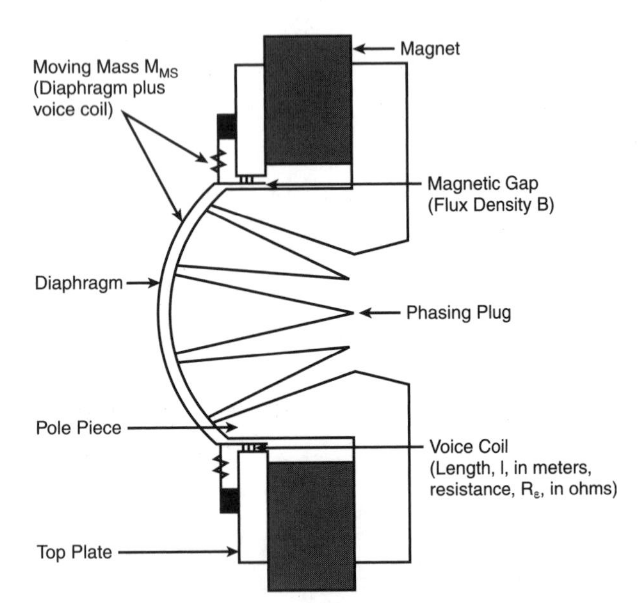

Figure 19.4 Diagram of a typical compression driver provided courtesy of JBL.

While a cone loudspeaker produces a sound field that is omnidirectional—going off in all directions—at low frequencies, a compression driver is usually attached to a horn, and the size and shape of the horn determine the coverage of the higher frequencies. This is an important distinction, as you will see when we start to look at box designs and deployment. Low frequencies tend to be omnidirectional. High frequencies are very directional, which means that the coverage of a typical box is defined mostly by the dispersion of higher frequencies.

Having said that, it is easy to fall into the trap of seeing the horn as being nothing more than a way to guide sound in a specific direction, which is not the case. Compression drivers are also often referred to as *horn drivers*. As the vibrations move from the narrow throat of the horn through its flare and to the exit, larger and larger "slices" of air are excited, which results in a much louder sound than was present at the driver itself—which is why we are dealing with horns in the driver section rather than in the "Box Design" section, where they would seem to be more appropriate.

Horns are most often used with mid and high frequencies because their shorter wavelengths lend themselves to this format. When you get down into bass frequencies, the amount of space needed to make a horn that properly transmits longer wavelengths is pretty prohibitive. One way around this is the *folded horn,* which we will cover in the "Box Design" section later in this chapter.

It is not only the size of the horn, but also its shape that contributes to the overall frequency response and to the dispersion pattern. And many horn designs have been employed over the years in search of one that puts out the most accurate sound and allows for pattern control. When it comes to pattern control, the idea is to keep the horizontal dispersion wide and limit the vertical dispersion for two reasons. First, most of the energy on the top part of the vertical dispersion shoots over the heads of the audience and is wasted. Worse, in an enclosed environment, those sound waves bounce off the ceiling and walls and back into the intended sound field, which results in muddy, incoherent sound. Second, because horns are usually in the same box or stacked on top of a cone driver, the two sound sources can interfere with each other, something we will look at in the "Box Design" section.

Most of the horns you run into will be either radial or constant directivity designs. A radial horn has two surfaces based on an exponential flare rate and two straight walls. While this works great for overall pattern control, the dispersion pattern narrows as the frequency increases. The overall effect is of a "beam" of high-frequency sound on axis, with those frequencies falling off rapidly as the listener moves off axis, resulting in a dull or muddy sound.

In May 1975, some of the beaming problems were addressed by D. Broadus "Don" Keele, Jr. of Electro-Voice. Instead of the typical practice of making a horn that flared at a constant rate from the driver to the exit point, by having a conical section in the middle part of the horn and then going into a rapid flare at the end, he was able to minimize the "beaming" of high frequencies, resulting in a much smoother response over a wider area. This design is referred to as a *constant directivity,* or CD, horn, and variations on it are still in use today. See Figure 19.5.

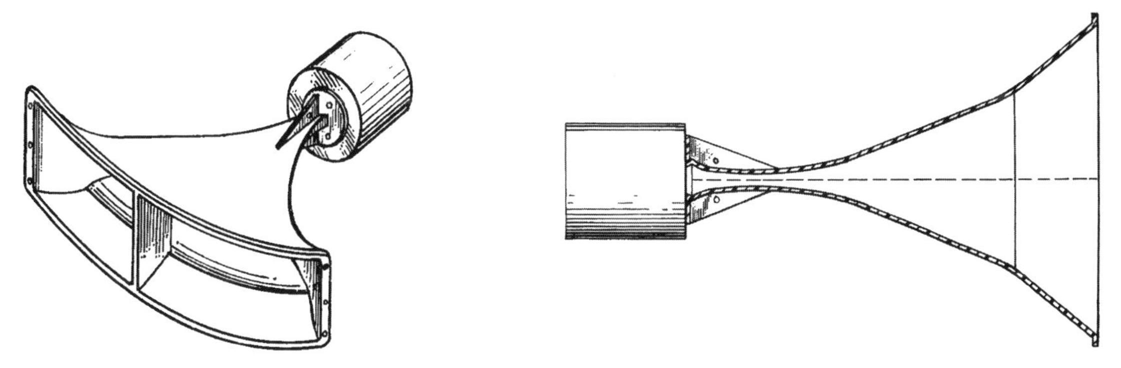

Figure 19.5 Diagrams from the original patent application for the constant directivity horn design, provided courtesy of Electro-Voice.

Most horns you will see are going to be of the radial or CD design or some variation on those designs, including Bi-Radial horns pioneered by JBL.

The last horn to look at is the multi-entry horn first patented by Ralph Heinz of Renkus-Heinz and also used extensively by Tom Danley in work at Sound Physics Labs, Yorkville, and Danley Sound Labs.

A multi-entry horn has multiple midrange and high-frequency drivers mounted to the same horn, as shown in Figure 19.6. The result is that more linear transient response and smoother polar patterns are possible, and greater power output can be achieved from a smaller box.

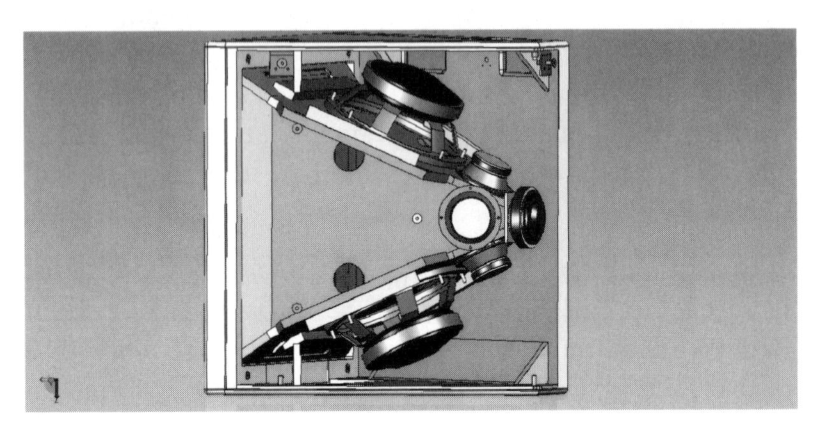

Figure 19.6 Multi-entry horn.

Impedance

Everyone repeat after me—oooooommmmmmmm. Relax your mind and take a chill pill—time to get into a little math....

Impedance is just what it sounds like. It is the degree to which the flow of electricity in any wire or circuit is blocked or impeded. Imagine this: You are holding a garden hose that has water flowing through it. Now imagine putting your thumb over the hose opening. If you cover the entire opening, you can actually still feel the pressure of the water, but nothing comes out of the hose. As you move your thumb, water is able to leave the hose, and the less you cover the opening, the lower the pressure behind your thumb. Think of your thumb as a kind of resistor and the degree to which you cover the hose exit as impedance.

Okay, here's the math part. Impedance is measured in ohms. (Get it? Ohms? Like the meditation sound I referred to above? Gosh, I'm witty....) Way back in 1827, a German physicist by the name of Georg Ohm put forth the idea that in electrical circuits, the current through a conductor between two points is directly proportional to the voltage across the two points and inversely proportional to the resistance between them, provided that the temperature remains constant. This is known as *Ohm's Law,* and it is expressed as voltage/impedance = current or voltage/current = impedance.

For example, if a speaker rated at 8 ohms receives a 10-volt signal, then the current is 1.25 amperes. If you lower the impedance of the speaker to 4 ohms, you double the current flowing to 2.5 amps. So, on the surface it would appear that the lower the impedance of the speaker, the more current (or power) is running through the speaker. And that is a good thing, right?

Not exactly, because as the impedance approaches zero, you are asking the power amp to produce more and more current, and current produces heat. More current means more heat, and you can easily toast an amp by asking it to produce too much current.

A couple of other things to wrap your brain around... (And yes, we will have a nap break after this chapter so you can recover.) First, the impedance of a speaker is dynamic. In other words, it changes constantly in relation to the frequency of the signal it is receiving. This is why speaker specs are listed as *nominal impedance.* It is kind of an average or best guess of the impedance at any given time.

Here is the other thing that is going to sound backwards: As you add speakers to a circuit, the impedance goes down, not up. How is that possible? Remember that the voltage output of an amplifier is constant. Adding one 8-ohm speaker to the chain requires a 10-volt amplifier to produce 1.25 amps of current. Adding another speaker that requires another 1.25 amps means that the same amplifier now has to produce 2.5 amps. 10/2.5 = 4. So now your total impedance is 4 ohms. Adding a third 8-ohm speaker that requires 1.25 amps brings the power draw on the amplifier to 3.75 amps. 10/3.75 = 2.6 amps. Remember, Ohm's Law also says that impedance is equal to voltage/current, so on that last example, 10/2.66 = 3.75 ohms. The bottom line for a guesstimate on total load is that every time you double the number of drivers all at the same impedance, the total load as measured in ohms is cut by half.

Just to make sure you were paying attention in math class, what if you have two speaker cabinets, one with an 8-ohm impedance and one with a 4-ohm impedance? Assuming that 10-volt amplifier again, the 8-ohm cabinet will require 1.25 amps of current, and the 4-ohm cabinet will need 2.5 amps. So the total power draw for the two cabinets is 3.75 amps. Voltage/current = impedance, so that means our two cabinets are presenting a load of 2.66 ohms (10 volts/3.75 amps = 2.66 ohms).

Power Handling and Efficiency

A speaker will have a power rating. This will usually be expressed as an average amount of power that the speaker needs to be driven properly. The unit of measurement is watts. (Yes, it's named after another physicist. James Watt was a Scottish engineer who lived in the 18th and early 19th centuries. He had nothing to do with electricity, but the unit of measurement was named for him to acknowledge his contributions to the development of the steam engine.)

Wattage is figured with—yea!—another mathematical formula. In this case it is the output voltage squared divided by the impedance. So, going back to our 10-volt amplifier and 8-ohm speaker, the power handling of the speaker would be 100 divided by 8 (100/8), or 12.5 watts. Modern amps put out much higher voltage, so using 60 volts and that 8-ohm speaker ($60^2/8$) gives you a power rating of 450 watts.

So a speaker rated at 450 watts should not be hooked up to an amplifier that puts out more than 450 watts of power, right? Again, it makes sense on the surface, but that 450 watts is an average, so a 450-watt amp hooked to a 450-watt speaker means the amp is being asked to put out its maximum power all the time. This is a good way to blow an amp. A good rule of thumb is that the amp output should be about double the power rating of the speaker.

Not only will pushing the amp to its maximum put the amp in jeopardy, it is also bad for the speaker.

Speakers fail in one of two ways: thermal failure or mechanical failure. Mechanical failure is always related to the movement of the voice coil. Too much movement in either direction results in what is called *over excursion.* If this happens as the coil is moving away from the magnet, the coil can "jump the gap" and either short out or cause the coil wiring to break, which results in an "open" coil. The bottom line in either case is that the coil no longer functions as an electromagnet, and the speaker stops working. If it happens as the coil is moving toward the magnet, the coil can bottom out, causing it to deform and be unable to move within the air gap. Again, no movement of the coil means no sound.

Thermal failure has more possible causes, the most common of which are too much input power, signal power outside the frequency range that the speaker can produce, or amplifier clipping. The first two of these make sense right away. Too much input or input frequencies that the speaker can't handle have to go somewhere. (Remember, energy cannot be created or destroyed, but its form can change.) In these cases the excess or unusable input gets converted to heat, which can deform the coil or actually make the speaker catch fire.

The last one is not so obvious. A clipped signal from the amp means that the amp is being pushed harder than it was made to. Since it does not have enough power to reproduce the entire signal, the top and bottom of each wave are cut off or clipped. Trying to reproduce a clipped signal produces heat, which can toast your speaker. Bottom line: Power matching is crucial, as an amp that is either too small or too large can result in a blown speaker.

One last piece of math before we move away from drivers and on to actual cabinets...

Sound output is measured as *sound pressure level,* or *SPL,* which is expressed in decibels and is arrived at using a standard of 1 watt of input measured at 1 meter of distance. Without getting really deep into the math, just remember that the amount of amp power needed to produce a specific amount of SPL doubles every time the speaker efficiency drops by 3 dB. So if a speaker with an efficiency rating of 93 dB puts out 120 dB SPL at

1 watt/1 meter, then a speaker with a 90-dB efficiency will require twice the power to achieve the same output level.

The rub is that as speaker efficiency goes up, the accuracy or fidelity of the reproduction often goes down, so achieving really good sound requires a lot more power. In the early (or even just *earlier,* as in prior to a decade ago) days of live sound, power was a precious commodity, so there was a constant balancing act between making speakers that were efficient enough to put out sufficient SPL and making ones that still sounded good. In the past decade, the output power of typical power amps has increased massively, which makes the efficiency of the speaker less crucial. It is, however, something you should still be aware of.

Time to move on...

Box Design

As you have probably already figured out, box design is not just putting together some wood and throwing a couple of drivers into it. The truth is that there is at least as much math and physics involved in good box design as there is in driver design. And it is not just design; material used in construction also has an impact on the final sound you hear.

If you really want to get into the nuts and bolts and calculus of speaker boxes, there are plenty of online resources to turn to. I am going to stay away from the math and go with an overview of the most common types of speaker design.

Sealed Box

This is not a design you will see often in PA systems, with the exception of some midrange drivers, but it is the most logical place to start. Think of the typical 4×10-inch half-stack cabinets used by guitar players. Now visualize a cone speaker in action. The movement of the cone moves or excites the air particles in front of it, resulting in audible sound. But what about the air in the cabinet? As the speaker cone moves, that air is moved as well, so what happens to it?

In a totally sealed cabinet design, the air vibrates the actual walls of the box, which is where that energy goes. But it takes a lot more energy to move a 1/2-inch-thick piece of wood than it does to move a paper cone, so the vibrations produced are not enough to really contribute to overall volume. And even though most sealed cabinets are not really sealed (they have a small air leak built in to allow for the equalizing of air pressure inside and outside the box), the box itself still vibrates, which can have a big effect on the timbre or sound quality.

One way to use the energy from the back of the speaker *and* keep the box sealed is to use what is called a *passive radiator*. This is basically a speaker with no signal routed to it. Instead of the cone moving in response to the signal being sent to the voice coil, it is moved by the air in the cabinet. This is not a common design in PA cabinets, but I have seen it in some studio monitors.

Bass Reflex

This is probably the most common design you will see. In a bass reflex design, there are holes, or *ports,* usually located on the same surface on which the speaker is mounted. It is one thing to just drill holes into a speaker cabinet, but it is quite another to properly design and build a port calculated to increase the speaker output at a specific frequency range. A bass reflex cabinet can have some internal structure attached to the port, which is calculated or tuned to maximize the output of the desired frequencies.

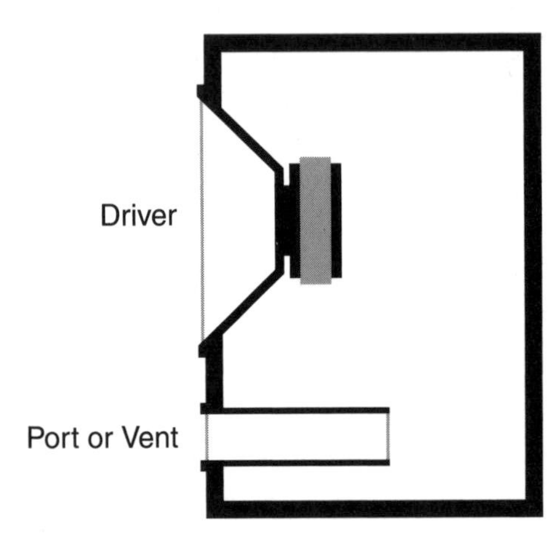

Figure 19.7 Typical bass-reflex enclosure design.

Because the "inside" energy is actually making it out into the airspace that the listener hears, bass reflex cabinets loaded with the same drivers and getting the same signal at the same power level as a sealed cabinet will generate more bass output.

Horn-Loaded

This term refers to any speaker box that uses some form of horn to combine the signals of a single or multiple drivers to increase the efficiency and control the dispersion pattern of the entire box. This can mean anything from drivers attached to a horn and the whole thing mounted to the box, to a design in which the box itself acts as a horn.

This last one is not common anymore (although it was very popular at one point in the development of live audio reproduction) and can take the form of a *folded horn.* This can mean an enclosure in which the driver fires out the front, and the back energy expands through some twists and turns to replicate the length of a horn needed to reproduce low frequencies, and those frequencies exit via a horn mouth, strengthening the output of the actual driver.

In a folded horn design, it is not uncommon to see only the mouth and not the actual driver from the outside of the enclosure. This is because the driver is mounted *inside* the box, expands through the internal horn structure, and sound exits from the horn mouth only.

Deployment

In all but small gigs, you will likely use multiple boxes of the same type, and how you place the boxes can radically affect the sound. Here comes the whole physics thing again. . . .

Each box can be termed a *point source,* which is the point at which a radiating pattern begins. This could be anything from a speaker to a light bulb to a rock thrown into a pond. In reality, each of the components in a box is its own point source, but we'll get into that in a minute.

Look at the diagram in Figure 19.8 and imagine the rock-into-the-pond thing. The areas where the "ripples" from each rock intersect are called an *interference pattern.* And it does not matter whether we are talking sound or light or ripples, the principle is the same. The point in the development of the waveform determines what will happen when the two waves meet. If they meet at the peak or trough of the wave, the waves will combine and become stronger. If they meet at some point in the middle, they will destroy each other and basically cancel out that part of the wave at that spot.

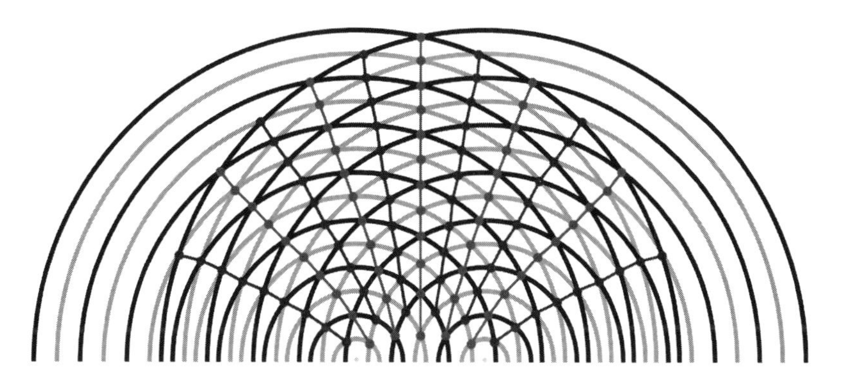

Figure 19.8 Diagram of two waves colliding. Illustration by Erin Evans.

In sound, the overall effect is called *comb filtering,* and it looks like what you see in Figure 19.9.

The audible result is a carved-out or hollow sound and much less overall output level. With multiple boxes, we need to place them in a way that minimizes this interference.

Many full-range boxes made since the '80s use a trapezoidal design that is calculated to minimize interference between boxes placed side by side. If you are using older boxes where all of the sides are at right angles, then you need to space them to avoid this kind of filtering of the sound. A friend of mine, Paul Overson, who has a small sound company in Utah, goes out on gigs with a homemade "template" that, when placed between his subs and top boxes, shows him the ideal angle at which to place his top boxes in order to minimize comb filtering.

One approach to multiple boxes that has become very widely used (although poorly understood by many sound personnel) is the vertical line array. The idea is that the

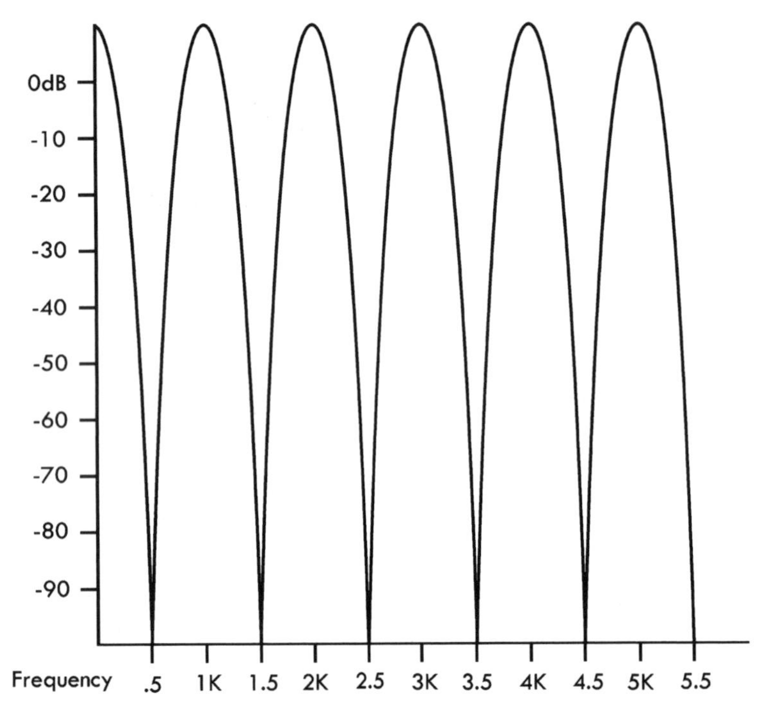

Figure 19.9 Comb filtering. Illustration by Erin Evans.

waveforms produced by a vertical stack or array of drivers of the same size will couple as long as the individual drivers are no farther apart from center point to center point than the size of the wavelength of sound at the frequencies being reproduced. The result is that the entire array acts like one big source with wide horizontal coverage and narrow vertical coverage, resulting in much clearer sound.

I am not going to get into the ins and outs of line array theory except to note that one of the advantages of a properly designed and deployed line array is that instead of output dropping by 6 dB with every doubling of distance, it drops only 3 dB. The bottom line is that you can get more sound farther from the source with fewer boxes.

As if just getting speaker boxes placed right on the ground were not hard enough, since the early '70s, the demands to minimize blocked sightlines have resulted in the practice of *flying*, or hanging, speakers. (Actually, the first flying systems were platforms that held speakers arrayed just like they would have been on the ground, but then the entire platform was lifted into the air. Bruce Jackson working with Elvis Presley and Stan Miller working with Neil Diamond were two pioneers of this practice.)

When flying speakers, good sound is complicated by serious safety issues. This is not a tome on the art of rigging, but just keep in mind that you need to really know what you are doing before you start hanging heavy speaker enclosures above an audience. *Never* put eye bolts or anything like that into a cabinet not designed to be flown. Speakers designed with flying in mind will have "flyware" built into them for this purpose.

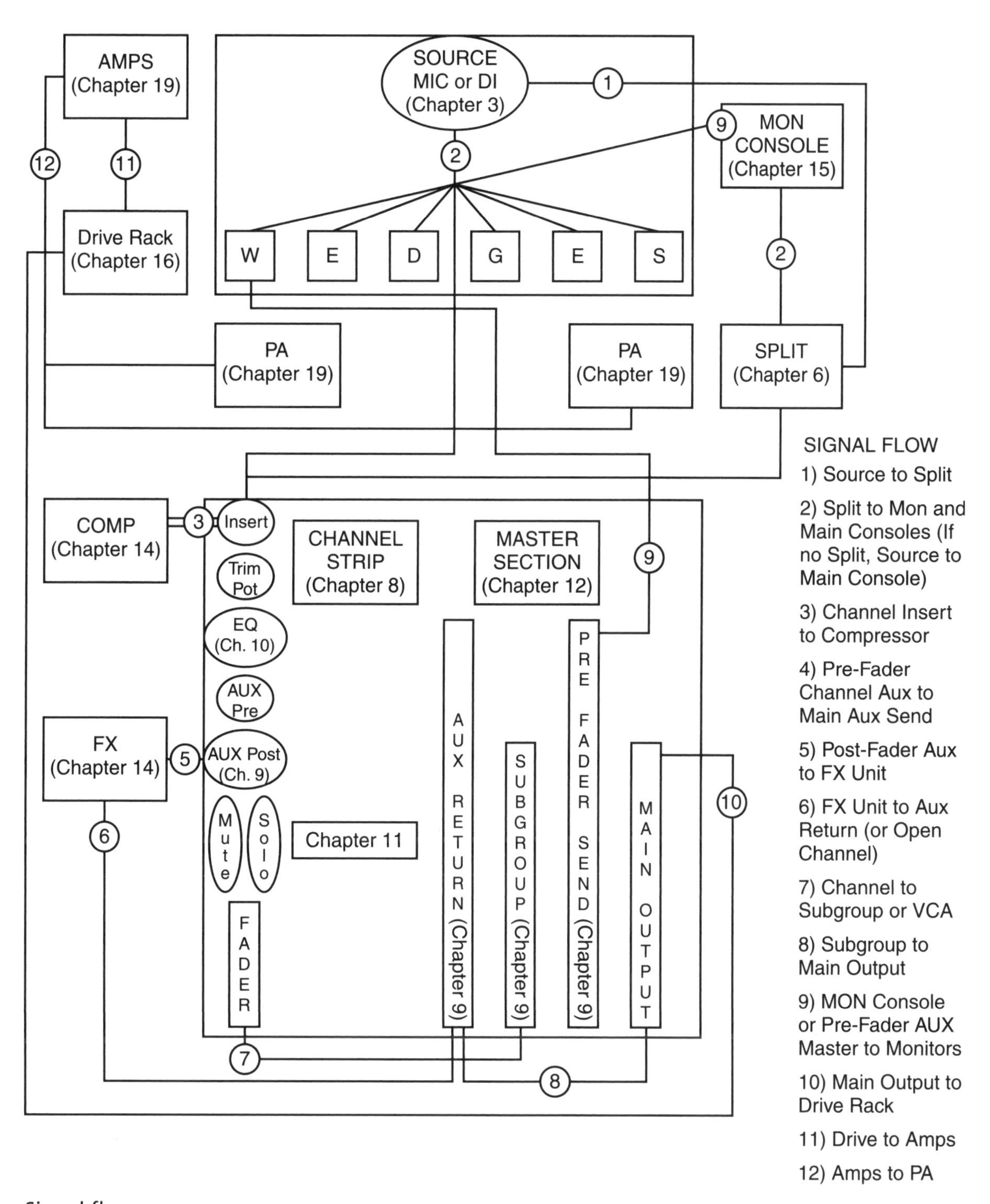
AMPS (Chapter 19)
SOURCE MIC or DI (Chapter 3)
MON CONSOLE (Chapter 15)
Drive Rack (Chapter 16)
W E D G E S
PA (Chapter 19)
PA (Chapter 19)
SPLIT (Chapter 6)
COMP (Chapter 14)
Insert
Trim Pot
EQ (Ch. 10)
AUX Pre
AUX Post (Ch. 9)
Mute
Solo
FADER
CHANNEL STRIP (Chapter 8)
MASTER SECTION (Chapter 12)
FX (Chapter 14)
Chapter 11
AUX RETURN (Chapter 9)
SUBGROUP (Chapter 9)
PRE FADER SEND (Chapter 9)
MAIN OUTPUT
SIGNAL FLOW
1) Source to Split
2) Split to Mon and Main Consoles (If no Split, Source to Main Console)
3) Channel Insert to Compressor
4) Pre-Fader Channel Aux to Main Aux Send
5) Post-Fader Aux to FX Unit
6) FX Unit to Aux Return (or Open Channel)
7) Channel to Subgroup or VCA
8) Subgroup to Main Output
9) MON Console or Pre-Fader AUX Master to Monitors
10) Main Output to Drive Rack
11) Drive to Amps
12) Amps to PA

Signal flow.

PART II

The Gig

20 Getting Your Hands Dirty

Now we have been through the entire signal chain, and you should have at least a basic understanding of the gear, how it works, and how each thing gets hooked up. So now it is time to start talking about putting together a mix.

Well, hang on there, little campers. There is still much to be done before we start messing with all the knobs and buttons. First, let's get real. The average touring or venue sound crew has from three to ten people, and only one or two of those is actually mixing. So what is everyone else doing? This is important, because if you are just coming into the business from school or some other training program, you will start out as one of those "other guys." The only way you walk out of school and into a mixing position is by starting your own soundco, by having a rich uncle in the business, or by going to work for an act with whom you have a preexisting relationship.

The second two options are pretty rare, so my first advice is to get yourself a job with an established company or venue doing the "other guy" stuff for a paycheck. This is still a business in which experience counts for a lot and time has to be spent paying dues. In the meantime, hook up with a local artist, promoter, or venue, do mix gigs at the usually pitiful prevailing rate, and look at it as experience and a stepping stone. A combination of those two things will keep your mixing chops up, keep the rent paid, and give you solid, real-world education about how things work on real-live gigs.

Kicked by the Wind, Robbed by the Sleet...

So, back to the question of what the "other guys" do. The world of live audio is not standardized by any stretch of the imagination. So you may have to adjust some of the terms I use here for what is the norm in your part of the country. I live in Las Vegas, and the terminology for audio personnel here is very different than what it is in L.A.—at least among local companies. The terminology we are using is something that comes from the theatre world and is commonly used by the union that represents audio people—IATSE, the International Alliance of Theatrical Stage Employees.

Audio guys are generally divided into three categories and are often referred to as A1, A2, and A3. And, yes, the A means "audio," and the numbers denote the place in the pecking order.

An A1 is usually a mix engineer for the house but can also be a system tech who tunes and maintains the system on a tour or in a venue. An A2 is often a monitor engineer and may also be a system tech for either house or monitors, depending on the venue and gig. As you can see, there is some nebulousness and crossover between A1s and A2s.

An A3 is an audio stagehand, and if you are the new guy on a tour, in a venue, or working for a soundco or even a union gig, this is where you will likely start out.

Duties of an A3 can include everything from pinning the stage (running connections from individual mics and DIs to the split or stage box), to wiring speakers under the direction of the system tech, to pushing cases and coiling cable.

The Union This is a touchy subject, and one I am going to get a lot of grief for addressing, but here goes. IATSE is not held in very high esteem by many in the audio tribe. The attitude stems from the insularity of many locals (I know plenty of experienced guys in Vegas who can't get a union card because the local is more concerned about protecting longtime members than it is about bringing in new, better-educated blood), as well as the reputation of many locals for sending out people based more on seniority than on qualifications. On the other hand, many gigs require that you are an IATSE member before you can even be considered. The balancing act is to be *in* the union but not *of* the union. In other words, you need a card to work in many places but a union attitude will get you fired very quickly from many gigs.

Don't despair that you went through all this training only to do grunt work. First, there is a lot to learn as an A3 if you can keep your mouth shut and your eyes and ears open. Second, it is all about experience. I have been running sound since the 1970s, but because I don't make a living at it anymore or even do it every day, when I go out and work for a soundco in Vegas (which I still do), I go out as an A3. In other words, I know how to do everything on the stage or dealing with the system, but I would have to work my way into an A2 or A1 position. It's not that far from the situation you will be in when you first start working. I'm just a lot older.

The need to understand the basics of implementing and deploying sound systems, from connecting mics and speakers to knowing enough about how things work to choose the right tool for the job, can't be overstated. And most of that knowledge is not something you can get from a book.

Had My Head Stove In . . .

There are some programs out there that take a real-world approach to live audio and go way beyond just teaching students how to run a digital console and how to mix. One that stands out is Belmont University's program, which operates as an actual sound company doing events both on and off campus. Students learn everything about doing real gigs, from prepping and loading a system to getting paid and getting the gear back to the shop. There are others out there that do similar stuff, and if you are going to work in the live audio field, it would behoove you to seek out that kind of program and not a glorified recording program that takes a few weeks to cover live sound.

As I said before, it is virtually impossible to take this stuff *too* seriously. In my day gig as the editor of *Front of House* magazine, I talk to mix engineers and system techs for the biggest tours and shows in the world and speak regularly to the owners of sound companies, from the small local guy to the enormous international touring firm with dozens of clients on the road at any given time. Not long ago, as I was putting together the outline for this book, I started asking these people what they wished green sound guys coming out of school knew when they got to the gig. Not a single answer concerned anything technical, with the exception of a few people who expressed the wish that new guys had a better handle on the proper mic to use for a given application. The rest of the answers were all about work ethic, teachability, and attitude. Some of the answers to the question "What do you wish they knew" appear near the end of the book. But the bottom line is very simple: Soundcos, whether they are the touring, one-off, or in-house venue variety, want people who are willing to work hard, not complain about it, and be open to suggestions and able to follow directions.

One of the biggest issues with new sound guys is a tendency to think they know everything there is to know when they arrive at their first gig. Worse, many will ignore or even actively contradict the instruction or advice of seasoned audio guys. Part of it is the arrogance of youth. Put it aside. The guy with the gray hair who gets a little lost navigating the latest digital console or system controller has a wealth of information and experience. And most of them are quite willing to share if given the chance.

But I'm Still on My Feet...

This is not just a question of the best and fastest way to get a job done—although that is crucial information when you are working a show that has three hours from the time the trucks arrive at the venue until things have to be ready for sound check. Many times this could be information that saves your life.

Working in a live production environment can be inherently dangerous. Think about it. We are moving and deploying big, heavy pieces of gear, large portions of which "fly" above the stage or audience. We work with electricity, and a large show may have things such as hydraulics and pyro to worry about as well. Every year we lose at least a handful of usually young techs because they were not following basic safety requirements. Experienced crewmembers only become such by keeping from getting injured or killed on the job. If someone tells you to do something that you know is cutting corners in the area of safety, you should refuse to do it and talk to a crew chief or production manager. But the instances of someone doing that will be pretty rare. This is a place where you really want to watch and listen to crewmembers who have some time in the field. This is a business where the best training is of the on-the-job variety.

Now it's time to get down to business and look at working a typical gig. For the purposes of this book, I will use a smaller gig with a crew of two or three people as an example. If you are on a big show, you will only do a fraction of this. But just as I used an old-school analog mixer for the examples in the gear part of the book because the principles are the same as you get higher tech and more digital, so it is with small-gig activities having a comparable function on bigger shows and in bigger venues.

21 Advance and Prep

The gig does not start when the band starts to play or even when you start setting up gear. The truth is, it starts weeks earlier in most cases. As we address this subject area, I'm assuming that you are going out as the audio provider. Either you are working for a small company or you are working for yourself. On a tour or working directly for a venue, these are not things you will be concerned with. But I find that even guys who work for a larger sound company often have their own smaller company as well, if only because that is the way they get to actually mix and do more than just what the crew chief tells them to do. And virtually every sound company I know started like this—from the smallest local shop to the biggest international touring company. Not everyone reading this will opt to run his own sound company. But I can virtually guarantee that most of you will at least try it.

As we move through the day on a typical gig, I am using as a point of reference a festival gig that I did for many years. This was a parish festival at a local Catholic church. The two-day event included everything from acoustic duos, to three-piece punk bands, to student big bands, to children's choirs and "folklorico" dance troupes, to a headlining dance band each day. Between acts we played canned music that I provided. Over the years I went from basic sound, to sound and lights, to sound and lights and my band as one of the headliners. The person at the parish I worked with went on to run a bigger fundraiser at the large private high school in town and hired me again. That was big enough that I had to hook up with another sound provider to have enough gear, and I got to mix one of my favorite singers from the '70s, Brenton Wood. (As a side note, one of the festival organizers ended up in my band and has been singing with us for almost a decade now.)

Here is the key thing: I got the gig when a sax player I worked with who had done the gig the year before called and asked me if I could take the gig because he just didn't want it. He told me that it was a giant pain and that the people were difficult and yada yada yada.... I ended up doing the gig eight out of the next ten years, and the two years I didn't do it were one when it was cancelled because of construction at the church and another when a new group of organizers decided to do it with all volunteer providers. They called me back for the gig the next year even though the gig was near L.A. and by that time I had moved to Las Vegas.

What I am getting at is that a gig that someone else decided was not worth the time led to lots of other work, and by the time we were done, I was making more than double what they were charged the first year. As time went on, my gear got better as the festival got bigger, but I did not keep the gig because of my gear. I got asked back over and over because I was easy to work with, did a good job, and went the extra mile to ensure that the festival was without hiccups. But most important was that I understood who I was working for.

Too often on gigs like this, local sound providers make the mistake of trying to act like a touring provider who is working for the band. Always remember that the person who is signing the check is the person who you are ultimately working for.

Example: One year the festival moved from its normal location to a pavilion outside the Rose Bowl. If you have ever seen the area around that world-famous stadium, you know that it is surrounded by multimillion-dollar homes. These people bought big, expensive houses up the hill from a stadium that holds about 100,000 people and expected that the stadium would never be used. And because they have money, they have power, and so the volume restrictions at the Rose Bowl are flat-out crazy. (Trust me, in another professional lifetime I was a newspaper editor in Pasadena, California, and the fight between the residents and people using the stadium for anything other than occasional football games was legendary and something I became very well acquainted with.)

I knew from years past that we were in for a "situation" because the crowd at this gig wanted to party, and the gig often went an hour past the scheduled time—which was fine with me, because I got paid extra. But the Rose Bowl rules said that any measurable amplified sound after 10 p.m. resulted in fines of something like $1,000 a minute. So I reminded the band of that when they set up. I reminded them again as they were starting their last set and again at 15 minutes out. And again at 10 minutes. And again at five, four, three, two, one... and at 10 p.m. I shut down the PA mid-song. The band was angry and made sure I knew about it. But the city sound cop went away happy, as did the organizer, and I got asked back again the next year.

Okay, lecture over. Back to the gig...

The first thing any sound person running a gig of any size should do is find out just what he is in for. This is known as *advancing the gig,* and it includes everything from communicating with the acts or their reps to determine what they need and making sure you can provide it, to actually going to the gig site to make sure you have the lay of the land.

Advancing the Gig

In a perfect world, you would be able to go to the venue a couple weeks before the gig, but this is not always practical, especially in a touring situation or even if it is a local act playing a one-off out of town. Regardless, there are certain things you need to know.

If you can't go and look with your own eyes, then make sure to ask all of these questions and get photos if you can.

- **Find out the venue location.** This needs to be totally specific, as in a real address and ZIP code. That last one is crucial. Why? I'll use a local example. There is a freeway called the 215 Beltway that circles the entire Las Vegas valley. So if the venue manager tells you the room is at "Durango and the 215," it could be in one of two very different places.

  Because the 215 is a beltway, it crosses Durango at both the south and north ends of the valley. And if you go to the wrong one, you will add a good hour to your travel time once you figure out you are in the wrong place, get directions to the right place, and make your way the nearly 20 miles between the two points. Or worse, suppose the venue verbally told you that it's at the 15 and the 215. Same deal: The two highways cross at two points again, and if you go to the wrong one, you may have more serious problems. Not only does the resort corridor known as the Strip lie between the two points, but so does the crazy interchange in downtown Las Vegas called the Spaghetti Bowl and the big construction north of that. An hour is getting off easy. Not only that, but the two areas may as well be on different planets. At the south intersection, you are a couple miles south of the Strip in a very nice, upscale area, and the clubs here are the kinds of places where the sound crew will be expected to do the actual gig wearing something nice, just like the folks in the club. Farther north, out by Nellis Air Force Base, the area is a lot rougher, and your main need will be a cell phone and perhaps some personal protection. And no one cares what you wear.

- **Get a list of what gear the venue has in house.** I am an advocate of the "I would rather have it and not need it than need it and not have it" school of thought, so I tend to bring more gear than I need to. So even if I am told that the house supplies an installed house PA and all I need to do is plug my console into their system, I will probably bring speakers and amps anyway, just as a precaution. But when the truck is packed, the speakers will be the first thing on it so that if I get there and find out I don't need them, they can stay on the truck and save time and labor on load in and load out.

- **Get an accurate description—with a diagram or pictures—of the load in/out situation.** The difficulty of the load in will have a huge impact on how much pre-gig time you need. Is there a loading dock? Is there a ramp? Remember that a typical loading dock is not helpful if you are carrying gear in a van. A ramp is much more helpful in that situation.

  How about the distance and terrain from the gear drop point to the stage? I remember one gig I did for years where the load in was down a fairly narrow corridor that was lined with nice wood wainscoting, and we had to be very careful to get the gear in and out without marring that wood. It added a good 20 minutes to the load in.

Are there stairs? My tech editor reminded me of a gig he used to do—by himself—where the stage was 10 feet and a flight of stairs above the main floor. Amps, speakers, everything up the stairs with no help. Fun, eh?

- **Find out where the FOH position is in relation to the stage.** On smaller club gigs, do not be surprised if there is no FOH position and you are expected to mix from the side of the stage. If that is the case, find out whether there is an area that you can use as a mix position and make sure you are carrying cable covers—pulling 20 feet of gaff tape off a snake at the end of the night is no fun at all.
- **Find out about the electrical situation.** If this is a small gig, chances are that the person you are asking for that information will have no clue. This makes it doubly important that *you* know what you need. A little more math: If you know the voltage, you can calculate the power needed to drive your gear. The formula is watt/volts=amps (current). So an 1,100-watt power amp in a 110-volt circuit needs 10 amps. Amplifiers and lights pull the most current of anything on the stage. For a typical audio-only gig, you are generally safe asking for four 20-amp services.

  Again, it is likely that the person you are talking with will not have that information, so—either in a trip to the venue to scope things out or before load in—there are a couple of things you can do to figure out your power. The first one is cheap (in the $50 range at Home Depot)—it is called a *circuit sniffer*. You plug one piece into an outlet and use the other piece to pass over the circuit breakers in the service panel. When you are above the breaker that controls the outlet you are plugged into, the sniffer will light up. A little bit of moving things between plugs and some trial and error, and you will find out just how much current you have available and where.

  Unfortunately, few small venues were wired with the needs of audio and lighting in mind, and it is not unusual to find that all of the outlets in the stage area are on the same circuit, which could spell disaster. When faced with that kind of situation, you have a couple of choices. The first is plenty of good, heavy-duty extension cords (*not* the orange ones from Home Depot). The other option is a power distributor. For a long time, this was something you needed a licensed electrician to hook up, but lately Peavey has been making a great piece called the Distro that plugs into a 220-volt appliance outlet and provides 16 outlets on eight 20-amp services. See Figure 21.1.

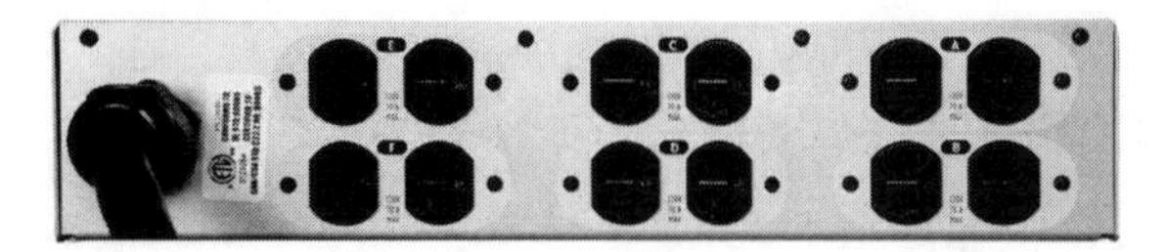

Figure 21.1 Back view of a Peavey Distro. Image courtesy of Peavey Electronics.

As long as you can get to a 220 outlet, this will solve most of your power issues. (Hint: You will usually find an outlet like this in the kitchen, so you will still need those good extension cords.)

- **Know who to call.** Make sure to get at least two names and cell phone numbers of people you can talk to if you get to the venue and something major has changed, the doors are locked, or whatever.

All of this is just advancing the physical venue—it is not the hardest and most detailed thing about advancing the gig. For that, we need to enter the magical world of tech riders and stage plots.

You may have heard or read about the outrageous demands made by some artists for how their dressing room is to be stocked. But this is not about bowls of M&Ms with the brown ones removed or fresh flowers on every table-like surface or a case of Jack Daniels in the fridge. There is a document called a *technical rider* that outlines the gear that an act expects. When you accept the gig, you are accepting the rider unless you have negotiated something different beforehand.

Riders can be really over the top in their requests. This seems to be especially true of up-and-coming artists. They may have been doing clubs on a 16-channel mixer and blown-out speakers six months ago, but now that they are really touring, they want the best. The list that follows is from an actual rider for a not-yet-famous touring act. Take a look at it, and we'll meet on the other side to talk about what it all means.

Front of House Audio System

(A) Front of house sound system shall be an active four-way stereo system, capable of producing an unequalized frequency response of +/−3 dB 50 Hz–18KHz at an undistorted signal of 120 dB SPL at the front of house console in any venue. For outdoor events, delay stacks should be made available.

(B) The FOH (Line Array Only) enclosures will be EAW, ADAMSON, V-DOSC, McCAULEY MLA-6, MEYER M3D, or JBL VERTEC. Any proprietary enclosures must be approved by Producer/Artist's Production Manager.

(C) The FOH speaker enclosures must be properly positioned and capable of producing a flat response for all sold seating areas. This includes front filled position (in front of stage) driven by a matrix or auxiliary send.

(D) The FOH speaker enclosures are to be powered adequately and accordingly to speaker and driver requirements.

(E) Power amps are to be CROWN, LAB GRUPPEN, or QSC. Ex: Crown IT-8000 to power subs. Crown IT-6000 to power mids. Crown IT-6000 to power mid-high and highs.

(F) Cross Overs/Systems Processors accepted are LAKE CONTOUR, XTA, KT, or BSS. (Located at FOH)

Front of House Console and Processing (Fly Dates Only)

(A) FOH console is to consist of fifty-six (56) channels. Ex: Consoles accepted: A. T. I. PARAGON II, YAMAHA PM 5000 or 4000, MIDAS XL-4, MIDAS HERITAGE 3000. NO EXCEPTIONS!

(B) FOH processing equipment is to consist of: 1. Lake Contour EQ, KT-Helix, KT-3600, KT-DN360, TC Electronics EQ Station, TC Electronics 1128, or BSS 1/3 Octave E.Q.'s.

(C) One (1) EVENTIDE H/3000, One (1) LEXICON 480, One (1) LEXICON 200, Two (2) YAMAHA SPX-2000, One (1) T.C. ELECTRONICS D-TWO or ROLAND SDE-330.

(D) Twenty (20) channels of compression, KT, BSS, or DRAWMER.

(E) Eight (8) channels of gates, KT, BSS, or DRAWMER.

(F) One (1) KT-DN60 or A.C.I. SA-3051 spectrum analyzer.

(G) One (1) pro compact disc player.

(H) Three (3) Clear-Comm stations with beacons and handheld sets. This is to be separate from lighting communications.

Monitor Console and Processing (Fly Dates Only)

(A) Monitor console is to consist of fifty-six (56) channels. Ex: Consoles accepted: YAMAHA PM-5D, YAMAHA PM 5000, MIDAS XL-4, MIDAS HERITAGE 3000. NO EXCEPTIONS!

(B) Monitor processing equipment is to consist of: 1. Sixteen (16) channels of E.Q. E.Q.'s accepted: T.C. Electronics EQ Station, T.C. ELECTRONICS 1128's (w/remote fader controller), KT-Helix, KT-3600, KT-DN360 or BSS 1/3 Octave E.Q.'s.

(C) Fourteen (14) channels of compression, KT, BSS, or DRAWMER.

(D) Eight (8) channels of gates, KT, BSS, or DRAWMER.

(E) Seven (7) YAMAHA SPX-2000, REV 500, or PRO R3 Reverbs.

(F) Five (5) Shure PSM-600 Hardwired IEM units.

(G) Eight (8) Shure PSM-700 IEM systems with beltpacks and antenna combiner.

(H) All necessary cabling for IEM systems and spare beltpacks.

Rider current as of February 2, 2010

Microphone and Mic-Stand Package (Fly Dates Only)

(A) Twelve (12) Radial J-48 direct boxes

(B) Four (4) Shure SM-58

(C) Four (4) Shure KSM-32

(D) Three (3) Shure KSM-27

(E) Five (5) Shure SM 57

(F) Four (4) Shure SM 98 w/ mic mounting hardware

(G) One (1) Shure Beta 52

(H) One (1) Shure SM 91

(I) Four (4) Shure KSM-137

(J) Twenty-four (24) round base mic stands with booms. Twelve (12) tall and twelve (12) short.

(K) Six (6) Z Bars.

Okay, let's start from the top.

- They want it loud.
- There are a lot of venues where a line array is a bad call, but they are spec'ing a line array "only." That along with the fact that one of the few speaker choices they are giving came out a decade before this rider was written and that the company in question has released at least three highly regarded line arrays since that time says one of a few things: 1) What they really care about is brand name. If you have one of those brands, you are probably going to be able to get the act's PM to sign off on it. 2) The rider was written by management and not the sound guys. 3) This young act took the rider from another act and just copied it.

Let's skip ahead a bit to the consoles. Note that the choices are pretty narrow and that the FOH models are all analog. The monitor choices include one industry-standard digital desk. Now go back to the list of possibilities of the origin of the rider and what that means. Given the console choices in conjunction with this, I can tell you virtually for certain that the act is traveling with their own FOH and MON engineers, and they either wrote or had a large hand in writing the rider. Either that or it is copied from another artist and those are the consoles *their* sound guys like.

So if you don't have all this stuff, does that mean you don't get the gig? No, it does not. In my experience the real make-or-break items are console, amps, and speakers. If I have at least two of the three of those, I am probably okay as long as one of those is the console. If you have all of the big three, then everything else can likely be negotiated. (Also note that this is a very strict rider. Most of the time you will see the words "or equivalent" next to every piece of requested gear.) The exceptions in this case are the mics and IEMs. Note that all of them are spec'd as being from the same manufacturer. What that usually means is that the artist or engineer is endorsed by said brand, which can throw a whole other layer of complexity into the mix.

The important thing is to balance the request against what you have available, and if the scale is even or tilting just a bit in your favor, then get on the phone and find out what is negotiable and what is not. You may have to put your sales hat on here—there are plenty of brands that work just as well as the ones listed, but you are going to have to make a case for this.

The Plot Thickens

The next items you will deal with are the input list and stage plot. These will vary radically depending on the kind of music the artist plays. For example, most rock and country acts will ask for a fully miked drum kit—that is, a mic on every drum (often two each for the kick and snare) and the hi-hat plus overheads. A jazz act may ask for just overheads and maybe a kick mic.

For our purposes, in Figure 21.2 I am using the plot I provide for gigs I play with my own band. Some terms to make sure you have straight: Stage left and stage right are always defined as being from the perspective of a person standing on the stage and facing the audience. In other words, if you are at the FOH position, stage left is always on your right. Got it? Downstage means closer to the audience.

First note that it is oriented as a bird's-eye view, and the text assumes that the person reading it is on the stage and looking toward the audience. Everywhere you see the word "vox" is a vocal mic. The drum miking is not specified because most of my gigs are in places where the kit is provided and pre-miked. Also, everyone uses in-ear personal monitors, so there are no wedges on this plot.

The input list may be an actual separate document with a list of the instruments and vocal mics needed, or it may be a series of notations on the plot itself.

The final piece of paper you need before the gig is the most important one—a contract. We are not going to get deeply into the contract except for a few notable items.

- List all of the gear and personnel you are providing.
- Note any substitutions from what is listed on the rider along with the name of the person who approved the change. When possible, make these changes via email so you have a paper trail in case of a problem later.
- Note the date and time of both the show and the load in/load out.

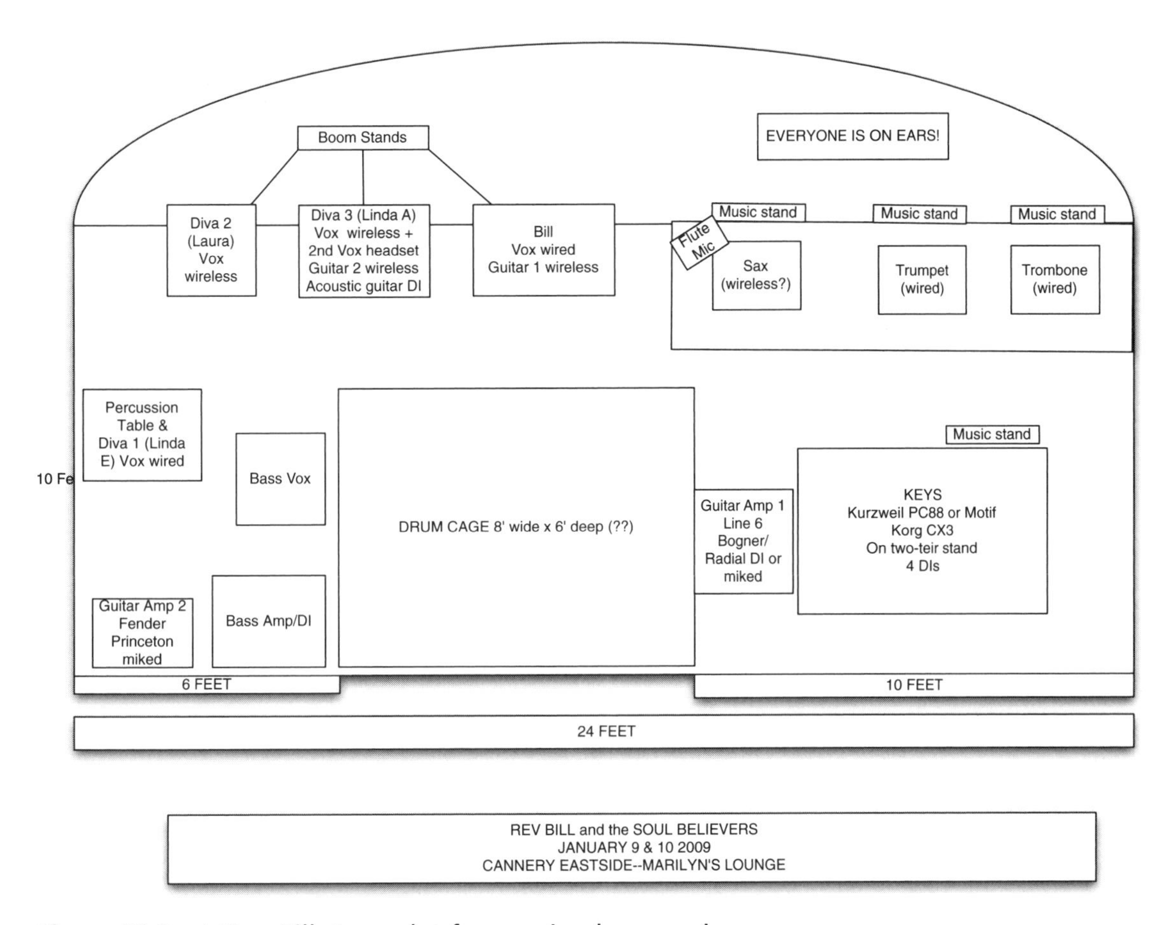

Figure 21.2 A Rev. Bill stage plot for a casino lounge gig.

- Note the amount you are to be paid, including a reasonable deposit, and the timing and form of payment. Unless this is someone you deal with regularly, it is pretty standard to ask for a deposit of up to 50 percent and to have the balance due and payable by credit card, cash, or cashier's check prior to the start of the gig.

The reason for the last part is that the sad truth is that local production companies are the first ones thrown under the bus if the gig goes badly. If the show sells poorly and the promoter and/or venue lose money, it is not your fault. But rest assured that someone will try to make it your fault, and you will find yourself chasing payment. Many sound companies make a policy of having the client/promoter pay after the system is producing sound but before the gig starts. Spell everything out and get paid in advance. After the gig, your only recourse is to sue. Before the gig, you can pull the plug and leave with the gear. The latter provides a lot more leverage.

22 On the Gig

Congratulations! You got the gig. Now what?

Again, we are looking at this with the assumption that you are working for a small company where everyone does everything, or you are working shows with your own gear.

Welcome to the Working Week

Most of the companies I know run on a Tuesday through Sunday workweek. The majority of gigs are on the weekends, so Monday tends to be everyone's day off, and a lot of small shops are just plain closed on Mondays. On Tuesday morning, the agenda often calls for breaking down and cleaning up gear from the gigs of the past week. Cables that went bad are repaired, rack cases get a fresh coat of paint, work boxes get restocked. In other words, everything that is independent of any specific gig gets taken care of.

Next up is looking at the gigs for the coming weekend and making decisions regarding gear and crew. Even if you are a one-man shop and there is one gig, this is a needed step. Time constraints, location, and even just the limitations of your body will often require hiring additional help at least for load in and load out. If the gig requires extra hands, then book them early—if you wait until late in the week, the good ones are already taken.

Side trip—I did a gig a few years ago at a county fair in rural Utah. The crew consisted of me and the sound company owner, who was mixing FOH. (I was mixing monitors.) The promoter was to supply three hands for load in and load out and to handle extra stuff during the gig. What we did not know at the time was that the labor pool consisted of inmates from the county jail. A good hand at least knows which end of an XLR cable to plug into a mic. Let's just say these guys were less than helpful. . . .

Okay, back to the shop, where it is Wednesday. Time to pull gear for the gig. If you are working for a small company, this may be a moot point because you will be using everything in the shop on every gig. But even if that is the case, this is when gear gets taken out of cases, inspected, and tested. Especially in cases where the gig is not near the shop, this is a crucial step. One thing you can have happen is that you get to the gig and have an equipment failure that could have been avoided with a bit of prep time in the shop.

If space and time allow, I like to take this time to set up the rig exactly like I will at the actual gig and test each part to make sure that not only are all of the individual parts working, but that they are all working together as well. If you have the means and the space, you want to test everything at as close to show volume as possible. There are problems that may not be apparent at lower volumes but that are very obvious at show levels. An intermittent connection on a piece of gear may not crop up in the shop at low volume, but under the assault of multiple subwoofers . . . well, you get the picture.

Once the rig has been assembled and tested, then it is time to pack it all back up and get it ready to go. Crucial: Always use a written list! The list will be your bible at several points—when packing up, when loading the truck, when loading into the venue, when loading out and repacking the truck, and finally when unloading the rig back at the shop. Use a checklist at each of these junctures, and you will have moved a long way toward a more successful gig.

Showtime—No Sleeping In for You

Show day has arrived, and it is going to start early and end late. Some people like to load the truck the night before the gig, but unless you have 24-hour armed security or the ability to park the truck inside a building and lock the building down, I do not suggest this approach. I almost never hear about gear being stolen from a shop, but I hear about loaded trucks and trailers being stolen on a pretty regular basis. Remember, theft is usually a crime of opportunity, and a loaded and unattended truck or trailer just screams "opportunity."

So how early do we start? Probably the most common reason for a gig going badly is rushing setup. Rushing leads to forgetting and shortcuts that come back far too often to bite you in the butt. The Rev.'s Rule of Thumb is take the amount of time you think it takes to unload the truck and load into the venue (of course, you already advanced the venue and are accounting for things such as the loading area being located farther away from the venue, needing to use freight elevators, stairs, and so on, right?) and double it. So if I think it takes two hours for me to load in, set up, and line check, then I shoot for being at the venue four hours before the band. The same rule goes for travel. If the venue says they are an hour from your shop, and you have checked Google Maps and they appear to be right, allow for two hours anyway.

So using that math, if the band is supposed to arrive at 5 p.m. for a 6 p.m. sound check with doors at 7 and the show at 8, and you have a two-hour setup and a one-hour drive, then I would suggest being at the shop by 9 a.m. Remember, you still have to load the truck.

Now we magically jump ahead in time. You are at the gig, the truck is unloaded, and the cases (or actual gear for things like speakers if you are not at the "case mandatory" point yet) are more or less where they need to be. Try to make sure that you have a place to store empty, or "dead," cases. If there is no place to do that (here comes the whole advancing the venue thing again . . .), and you have to put them back on the truck, then that time has to be figured into both load in and load out. The order in which setup occurs really depends on the type of gig, the number of people in the crew, and the way a particular group does things. If you are new, it's best to find out how things are done and then do them that way. No one wants to hear the new guy saying, "Well, when I worked for Rock Star X, we did it *this* way." No one cares how you used to do it. One of the cardinal rules of the gig (more of which you will find as we approach the end of this tome) is that there is the right way, the wrong way, and *our* way.

The big exception here is when the issue is safety-related. If you are being asked to do something that puts you or someone else in physical danger, then bring it up to your direct report. If they can't or won't address it, then take it up the ladder. If the answer you get is, "That's just the way we do it," then be prepared to bail and lose the gig and probably any future work with that company. But do it, because it is better to be gig-less than dead. There have been far too many instances of new crewmembers doing things in an unsafe manner either because they were told to do it that way or because they just did not ask and get told the right way. And too many young crewmembers are killed or crippled every year. Don't become a statistic. Safety is paramount—always.

How We Roll

My approach is to get the main part of the system set first: FOH console, speaker processors/crossovers, amp racks, and house speakers placed, wired, powered up, and tested. When I know that part of the system is up and working, then I can deal with everything else.

You may notice that I am not giving specific instructions on how to connect parts of the system. That is because although certain principles apply to every system, there are too many variables even with legacy analog systems to do an accurate overview, and the onset of digital makes it even more open to interpretation. But there are some inviolate rules, mostly having to do with power.

- Whenever possible before connecting two pieces of gear, make sure both are powered off.
- If you have to plug into something live—say, a line from the stage into a console channel—make sure the receiving end is powered down. In the case of a console, mute the channel.
- Mute the console before powering up or powering down any gear hooked into it. Many pieces of gear will produce a very loud transient on power up or down, and

that transient can damage amps and speakers farther down the line. Muted or not, never turn the console on or off when the amps are still on.

- When applying power, start at the "small" end of the signal chain and move toward where the signal is the loudest. On power down, go in the opposite direction. Stage elements (not including monitors) first, processing gear next, console (muted and main faders down), speaker processors, and finally power amps. Then unmute the console with the main faders still all the way down and slowly bring up the faders to avoid any nasty surprises.

When working with digital components, the cardinal rule is that once the signal is in the digital domain, leave it there as long as you can. Every A/D-D/A conversion degrades the quality of the audio. If you are using a digital split snake, and your console has the appropriate inputs, use them. The same goes for every step in the chain. Ideally, the switch from analog to digital will take place as close to the source as possible, and it will stay digital until as close to the speakers as you can get.

Getting Pinned

Once the main system is up, do the same thing for the monitor world. Once you know that everything is up and on and that signal is getting from the console to every wedge or personal monitor receiver and making sound, then you can move on to setting, or *pinning,* the stage.

Refer to your stage plot. If backline (amps, drums, and so on) are already there or you are providing them, then get everything miked and direct inputs plugged into the system. If backline is coming with the band, then set mics and DI where the plot indicates they should be. Line check every input at both mains and monitors. Have someone on the stage tap each mic as you confirm signal is getting to the console.

Once you know everything is working, it is time to dress the stage. This means making sure that cables are out of sight and do not pose a tripping hazard. Gaff tape (*never* duct tape), plastic cable tunnels, and stage draping are all your friends here. And although this step may have nothing to do with you providing good audio, it may have everything to do with you getting the gig next time.

When you interview for a job or meet with a potential client, you don't show up in shorts and a T-shirt, even if that is what you wear every day at the shop. If you're smart, you don't do the gig dressed like that, either. Think of the stage as an extension of you and your services. Given similar production experiences and prices between two companies, with one providing a stage that looks tight and clean and the other a stage with cables strung across it and scratched and dented gear, which one will get the call for the next gig?

With the stage set, pinned, and dressed and audio being produced by the system, you are off the clock until the artist or their rep arrives. This is a good time to get with the client and get paid. Unless this is a client with whom you have a regular, ongoing relationship, you need to do everything possible to get paid prior to the gig. If you are working for an entity like a corporation with invoicing and billing policies that prevent getting paid before the show begins, then make sure you have a hefty deposit. My policy is that I always look at any money that comes to me after the gig as gravy. I have to cover all costs, including labor and fuel, with the deposit. After the gig—unless it is a regular customer—the only way I can go after monies owed and not paid is through the court system, which can take a long time and can be less than effective. I once had a contractor working for me on some home repairs. Things were going poorly, and I finally lost it with him and told him that if I did not get what I had paid for, I would sue. He told me to go ahead, that he would just go out of business and reopen under another name, and that there was nothing I could do about it. The point here is that if someone is planning on cheating you, they'll find a way to cheat you. The only way out is to get paid in advance. A good contract can help, but given the choice of a good contract or cash in hand, I'll take the *dinero* every time.

23 Hello (Hello . . . Hello . . . Hello)—and Welcome to the World of Delay

The terms *delay* and *echo* are often interchanged. In the world of FX that is fine, but we are now talking about time-aligning a system, and the idea is to get rid of echoes, not to create them.

Without getting into scads of physics, there are a couple of reasons why this is a necessary step. Let's start with a situation where in order to achieve even coverage, we need to place speakers in places other than the main hang. Now let's say an audience member is standing next to one of those auxiliary speaker stacks. If you don't align the system, what that audience member gets is the sound from the speaker closest to him followed—often louder—by the sound from the main PA. Why does the sound from the speakers farthest from the stage hit *before* the sound from the speakers right at the stage? Because light moves much faster than sound, and the electrical signals in the system are moving at the speed of light. So, the signal to that second speaker placement gets to that speaker at virtually the same time as it does to the main speaker hang. The sounds from each source are produced at virtually the same time, and it takes the sound from the main stack longer to hit your ears than it does the sound from the speaker right next to you. Depending on the distance between the two, the result will be anything from a muddy, indistinct overall tone to actually hearing the same sound as two distinct events. And that is why those aux speakers are usually called *delay stacks*. To clean up the sound, you need to apply delay to the speakers farthest from the stage so that the sound emanating from them is formed at the same time the acoustic energy from the main stage is arriving at that location.

And it is not just big gigs with multiple speaker locations where this is an issue. In fact, small gigs often sound horrible because the PA has not been delayed to the backline. Brian Klijanowicz, who writes for me at *FOH* magazine did a great piece on this subject that he has allowed us to use here. Take it away, Brian . . . .

A Realistic Approach to Subwoofer Time Alignment Time alignment is a very important, yet very often overlooked aspect of system setup and tuning. A correctly time aligned system has many benefits, including more even coverage where two sound sources overlap and a more even response across acoustical crossover points. It can give even the cheapest of systems a couple decibels more in the area where engineers tend to like them most: bass frequencies. This leads to the topic for this article—subwoofer time alignment.

Two ways to quickly achieve this are by using a sine wave and by delaying the PA back to the kick drum. Either one or both will work better in different situations depending on the size of your gig, the time you want/have to work on it, and just how much you care.

With the more frequent use of digital crossovers, system controllers, amps, and consoles, it has become easier than ever to add delay to multiple signals, whereas years ago you would have to eat up an entire rack space and insert cabling for just one channel of delay.

This is intended to be a minimalistic, quick way to time align your system, so only a few pieces of gear will be needed. Using a sine wave will only require a sine wave generator and the capability to delay an output signal and to reverse the phase of it. To delay the PA back to the kick drum, all you need is a channel of delay and your "golden ears."

For this piece, we will assume that the sound system we are working with is a standard front-loaded stack configuration. This means there is a subwoofer (producing subfrequencies from roughly 100 Hz and down) ground-stacked on the floor with a top box (producing roughly 100 Hz and up) directly on top of it. We will also assume that both subs and tops are currently producing the same polarity and facing the same direction (azimuth/splay angle).

Sine Wave Usually an engineer would not want to use polarity to cancel out a signal. But that is the whole concept behind this technique.

First, it is important to find the acoustical crossover frequency between the subs and tops. A measurement device is the most accurate way to do this. But if you don't have one, there is an easy way to rough it in. Assuming you don't have a measurement device, flip through the pages of your crossover to see what the crossover frequency is and use that. (For this piece, it will be 100 Hz.)

Let's get back to the concept. We will take two similar sound sources, sub and top, that are playing a 100-Hz sine wave and flip one out of phase. The crossover frequency is the point at which both sound sources overlap and start to fade off from one another. But more importantly, they will still reproduce 100 Hz, because that is the beginning of the crossover filter on each source.

By physically looking at the speaker cabinets and knowing where the drivers are in the boxes themselves, you will be able to determine which signal to add delay to. Typically in the club world, top boxes are placed a little behind the front of the sub. That way if the top box falls, it will fall back and not onto the drunken audience. So, assuming this is the case, delay should be added to the sub. While the 100 Hz is playing through the sub and top, start increasing the delay to the subwoofer signal. Eventually, the two signals will start to cancel out, and the total SPL will reduce substantially. Find the point at which most cancellation occurs and leave it at that. Change the signal that's out of phase back into phase, and you should have summation at the crossover point.

Some crossovers have an on/off function for the delay. A good way to check whether summation is occurring is to flip the delay on and off to hear the difference.

This whole process can sometimes be done with music—preferably with driving, kick drum–heavy music. However, it can be hard to distinguish what is cancelling and summing in the crossover region with full-range music playing. A quick fix on a digital console would be to throw a low-pass filter on the music channel.

Back to the Kick Delaying the PA back to the kick drum is very simple. It works very well in the small "hole-in-the-wall" biker bar, but once you get into a bigger venue, it can become a bit harder to hear the difference. Smaller venues that have lots of reflective surfaces and 90-degree walls create standing waves. This makes it almost pointless to spend the time to get out a measurement device—especially when the club owner wants you to be done setting up before the first road case is even in the building.

Since this technique works better in small venues, we will assume your PA is in a small venue and set up in front of a band. Even though it's still surprising to club owners and patrons, we all know how loud a drum set can be in a small room. That is why, typically, most engineers won't even turn up all the drum mics, or they just won't mike the whole kit.

So the drums are roughly five feet behind the PA. Think of the kick drum as another speaker. If you are sitting in the audience, would you want to hear the main PA with another set of PA five feet behind it? Probably not. So if you are sitting in the audience listening to a band, you probably would not want to hear the kick drum once through the PA and then again five feet later.

Most drummers do relatively the same sound check routine. They will hit each drum individually with quarter notes at a medium pace until you ask them to switch to the next drum. When you get to the kick drum, get a rough sound in on the channel strip and start tweaking the delay back. Add a little bit of delay at a time. You will start to hear a change in tone and depth. Once you get to the sweet spot and are happy with the sound, that's it!

Thanks, Brian. Before we leave the whole subject of delay, there is a reason to mess around with delay besides cleaning up sound and getting even coverage. I am just going to touch on this because it gets into the whole idea of psychoacoustics, or the way humans *perceive* sound. This is done in the theatre world all the time. Think of the coverage area—especially the source point of the PA—as a kind of stage. Great care is taken to "place" individual voices and other sound sources in a way that results in the reinforced sound coming from the same direction as the original source. Then, the whole PA is delayed back so that it is just a few milliseconds behind the actors onstage. The conceit of theater is that real actors don't need a PA. That may have been true in the small acoustically designed theaters of times past. But with huge traveling productions of Broadway shows becoming the norm, and as more contemporary and rockin' music has made its way to the stage, those days have gone the way of $35 seats for a Broadway show. Really good theatrical sound designers know how to use placement and delay to fool the audience into believing that the sound they are hearing is coming from the stage and not from a PA system. Here is the psychoacoustic part: If a sound starts in one place and then continues from another source, the brain will think that the entire sound is coming from the initial source.

Get it? By making sure the initial attack is heard right from the stage and that the delay to the PA is an imperceptible few milliseconds, the brain will tell its master that the sound is coming from the stage. Neat trick.

24 Backline Basics

We are going to step away from the sound system for a minute and into another area that can be a great job for those into live event audio. You see those things on the stage that are making all the noise? The amps and drums and stuff? Well, here's something that may come as a bit of a revelation: Except on really small and really big gigs, that stuff is usually not carried by the band.

It comes down to a case of straight economics. At the low end, local bands and the proverbial "band in a van" are carrying everything they own, which may actually include basic PA. At the high end with big tours, they are carrying all backline and production (audio, lights, video, and so on). And there is business and money to be made everywhere in between. Some acts carry backline and board groups (consoles and processing) and rent "racks and stacks" (amps and speakers) locally. And plenty carry only things that they can get into the overhead bin on an airplane. The rest of it gets rented.

Behind the PA

This is the wonderful world of backline—pretty much everything behind the PA. It can include guitar and bass amps, drums, percussion rigs, keyboards, and drum kits. It can occasionally mean instruments such as guitars and basses (although they are usually there as backup, with the player bringing one main instrument on him and renting a backup).

So, two questions. First, why would you want to deal with this stuff? The answer to that is twofold and pretty easy: There is a lot of money to be made if you do it right. (One sound company owner I know got a nice Gretsch guitar as a Christmas present from the backline company he uses, which gives you an idea how much business he was sending to them.) The second part of the answer is that although there is generally less overall business in backline in any given city, the ratio of provider to amount of business works much more in the favor of the provider than the "real" world of audio does. The bottom line is that you make a lot less per gig than the company providing the big stage system is billing, but to use my adopted hometown as an example, on a given weekend there may be a dozen sound companies doing gigs in 20 venues. But 10 of the 12 companies are all using the same backline provider, and no one is angry about it.

The second question is whether you can do full audio and backline as well and just bill for the whole thing. The short answer is yes, but the truth is a lot more complicated. The two services require some very different investments and skill sets among those working the gig. It takes every bit as much knowledge to maintain and set up backline as it does to do the same with a PA. It is just a different kind of knowledge.

Getting the Gear Right

If you are thinking about doing the backline thing in addition to sound reinforcement, there are ways to do it and make it work. The biggest challenges are getting the right gear in and getting someone to care for it and set it up. Drums are pretty easy. Many—if not most—drummers carry their own cymbals as a matter of course or can be persuaded to carry them, so if you have good-quality "concert" kits with at least two mounted toms, two floor toms, and double kicks or a double kick pedal, and they say Yamaha, Pearl, Ludwig, or DW on them, you are probably okay. When setting rental prices, make sure to keep in mind that even though drummers use the same heads on their own drums for a significant amount of time, they will generally expect the heads on a rental kit to be new.

You or your backline person need to know how to change heads, tune, and set up a kit at minimum. But drums are pretty simple overall.

Percussion rigs are likewise pretty standard. An LP percussion table and a set of congas that includes a conga, a quinto, and a tumbao (three sizes of conga drums) with a Toca or Latin Percussion logo on them in a nice basic black will usually be sufficient. Percussionists will usually bring a suitcase full of hand percussion toys.

Guitar and bass rigs are simple. No stompboxes—most guys bring their own. For guitar, a good Fender Twin, a Marshall head and half-stack cabinet, and maybe a VOX AC-30, and you're in business. You or your tech needs to know how to do basic maintenance and repairs, plus how to change tubes and bias the amp correctly afterward.

For bass, an Ampeg SVT and maybe a Gallien-Krueger RB800 plus a single or dual 15-inch cabinet and a 4×10-inch, and you're set. You have the same maintenance and repair issues as with guitar rigs.

This hardly covers every request, but it will cover the vast majority of them. I once got a call from a local venue to rent my Line 6 Vetta guitar amp because the rider called for one and they could not find one at a rental house in town, but they knew I had one. The reason the rental houses did not have one is that people who want an amp like that will almost always carry one with them.

This brings us to the biggie: keyboards. This is the tough one, because remember that most synths are just big computers with piano-style keyboards on the front end. And your computer is obsolete by the time you get it from the store to your car, so what do you think happens with keyboards? Same deal. It becomes a case of making sure you have certain meat-and-potatoes stuff—Kurzweil PC88 piano, Yamaha Motif, Korg Triton, and a good Hammond organ (very expensive and notoriously hard to maintain)—and you will be able to cover most requests.

A word about organs: There are very good tone-wheel organ emulators out there made by Korg, Nord, Roland, and even Hammond but . . . I did the keyboard rental thing for a couple of years. I had a Motif, a PC88, and a really nice Korg CX-3 organ. The Motif and PC88 paid for themselves several times over. The CX-3 never did, despite the fact that lots of guys play them. The bottom line is that if the request is for a Hammond B3, then only a B3 will do. As for the rest of it, keeping up with the latest flavor and feature set is really hard and very expensive, and each one you buy is really just a bet on how the market will go over the next year. Stick to the basics.

25 Hands on the Knobs

We are finally going to put together a mix. And you would think this would be the longest, most detailed chapter in the book, but it is probably one of the shortest and most general.

Mixing is very subjective, which makes it vital that you know who you are mixing for. I can hear the collective sighs now: "The audience." That's the obvious answer. Except the audience does not sign the checks. If you are to succeed, one of the hardest skills you will learn is to identify who the person signing the checks is listening to when it comes to production matters, including the quality of your mix, and make that person an ally.

When it comes to the actual mix, I have found some general principles to be helpful.

Fix It at the Source

The old joke about "fixing it in the mix" is only funny because those who really know will tell you that there is only so much you can fix in the mix. Garbage in equals garbage out. A lousy guitar sound coming off the stage can only be fixed to a certain degree in the system.

It is always best to get the source sound as close to what you want to hear in the systemas you possibly can. Once again, here is where your interpersonal skills come into play. You can't go to a lead guitar player and start ripping apart his tone. A better approach is to say, "It sounds great onstage [regardless of whether it does], but I am having a really hard time getting it to sound like that in the system. If we could roll back the low end a little, then I can get more of your sound in the system, and we'll get the low end back to you onstage through the monitors."

That still may not work, but telling the guitarist his tone stinks is a guaranteed non-starter.

The other big stick in your arsenal is mic selection and placement, which is a subject that could fill a book all by itself. For our purposes, let's just say that you should experiment with different mics on different sources with different placement anytime you get the chance (but *not* on the gig).

But don't get so set in how you do things that you can't try something new or take something someone else does and apply it to your gig. Mic selection and placement is one of the places where there is no real substitute for experience. Listen. Keep your eyes open. Ask questions. And experiment—on your own time.

Find Your Foundation and Flow

How you set up a mix should be determined by the kind of gig, the kind of audience, the kind of music, and the gear you have to work with.

When I first started with live sound as a musician, the PA was just the way we got the vocals up above the band. Instruments and amps did not get miked unless they were horns or acoustic guitars. And we were using small boards, so I always put the lead vocal in Channel 1. And I kept doing things that way for a very long time.

I have since migrated to the typical rock setup of the kick in Channel 1 and the first group of channels dedicated to drums. But this only happened as I started to do gigs where everything was going through the system, because then it made sense. The kick, snare, and bass were the foundation upon which the music was built, so once I was past the stage of using the PA to add vocals to a mix of instruments coming off the stage, it made sense to start there.

The foundation of the music will change by genres. For example, a straight-ahead jazz drummer will almost never even have a mic on the kick. A couple of good overhead mics to pick up the entire kit is a common setup. The hi-hat becomes the rhythmic foundation, so on a jazz gig I might put that in Channel 1. Note that if you do all rock gigs with the kick in Channel 1, you make any assignment changes at your own peril. When an adjustment needs to be made in the heat of battle, instinct takes over, and if you reach for what you think is a kick mic and it's a hi-hat, it can be a problem.

A couple of weeks before writing this, I was out on a gig in an arena that was using a totally computer-driven mixing system with no physical control surface. Because the control surface was virtual, it could be adjusted to the mixing styles of individual engineers. And the two guys on the gig had very different styles and were able to fit the surface to their style, and neither had the inputs in anywhere close to the order that the other one had them in.

Finally, don't force a sound that is inappropriate to a musical style just because it is what you are used to. Folk music needs to be mixed very differently from a rock show. And failing to make that adjustment to honor the music is a huge mistake. I once heard Howard Paige (whose list of A-list clients takes up several pages) ask the immortal question, "Since when did the kick drum become the lead vocal?"

Listen to every kind of music you can find and have a solid foundation of what it is supposed to sound like before you take a gig mixing that genre. Then go with the flow and let the music do the talking. It is not about us or our huge systems and egos. It is about letting the artist onstage communicate his artistic vision to as many people as possible with finesse and clarity. (Unless it is a punk gig... just kidding!)

Triage

In medical terms, this is the practice of surveying the victims of some kind of accident with a lot of injuries and deciding who to treat first based on the severity of the injuries and the odds for survival.

In many situations today, we have the advantage of recorded shows and the ability to mix shows before the curtain ever rises. It is not unusual for an engineer to spend a week or more building a mix in a rehearsal studio using a recording of the act. But in one-off situations, you may be mixing an act you have never heard.

Spend some time *listening* before you start adjusting anything. Get your main elements under control. Drums and vocals are a great place to start. Then bring in what is needed to build on the drums and support the vocal. If there is something you can't fix, try to kill it in the mix if you can and move on. Remember, it is not just an order-of-importance thing; it is also a "Can I make it better?" thing.

And here is where knowing the music really comes into play again. Mixing horns is a great example. People who do not understand horn-driven music will often get the mix priorities all wrong. For example, in a three-piece section of trombone, sax, and trumpet, what gets the short end of the mixing stick? The trombone, almost all the time... But an engineer who has really listened to the music is likely to tell you that the 'bone is the most important voice in the section—the foundation, as it were—and that he builds the sax and trumpet on top of a really solid trombone sound.

Know the music. Know what's important. Know where you can make the biggest difference.

Less Is More—Softer Is Louder

This is especially true when mixing monitors, but it applies to FOH mixing as well. You can often make a bigger difference and a better mix by taking things out rather than continually boosting things. Here is that whole "know the music" thing again. Knowing the basic frequency range of various instruments is one of the most helpful pieces of knowledge you can ever possess.

Let's use monitors as an example. When a vocalist complains that he can't hear himself, don't give in to the immediate reflex of boosting that voice in the mix. Instead, take a good listen to the mix and identify things you can take out in order to carve out more sonic space for that vocal. Maybe it is not about boosting the vocal, but rather bringing down the guitar and giving the singer just kick drum instead of the entire drum kit in the mix.

At FOH, this really applies to EQing a specific input. Make it a goal to always cut and never boost EQ. No one can do that 100 percent of the time, but make that your goal. For example, consider a hi-hat that is not cutting through. Maybe dumping all of the frequencies below the high-mids is a better answer than boosting the highest frequencies. Or consider a guitar and vocal that are fighting each other. Maybe scooping some of the sonic range that the guitar and vocal share out of the guitar sound is a better approach than just pushing the vocal higher in the mix.

Keeping your mix subtractive rather than additive will keep your sound cleaner, will make it feel louder at a lower total volume level, and will lessen the possibility of feedback.

26 Touring Is Not for the Weak

There are many places where you may end up working in the sound biz—churches, schools, performing arts centers, theatres, resorts, and so on. But no matter where we end up, most of us at least toy with the idea of going out on a tour. There is a certain romance to the whole idea of being part of a big tour, and it can be a very cool thing. Plus, a major tour always looks good on a resume, no matter where you end up.

Ironically, living in the Live Entertainment Capital of the World (a.k.a. Las Vegas), most of the guys I see on big shows got there because they were looking for a way to get *off* the road. Once you are a touring guy, it can be hard to settle down and find something that pays as well and is as much fun as being on the road with a rock show. But the job itself is a lot harder than most people think.

Before you get on the bus, it would behoove you to have a good idea of what a typical touring day is like. At the time of this writing, one of the biggest shows out was Bon Jovi's "The Circle" tour. Mike Allison is a 30-year road veteran who works for Clair—one of the biggest sound companies in the world—and is the FOH system tech and audio crew chief on the tour. You might even say that touring is the family business, as Mike's son recently joined the tour and is part of the core audio crew that travels with the band.

Mike was kind enough to take time from his very busy gig to make notes on a typical day. I get tired just reading it. Want to go out on tour? Here is an idea of what your day might look like.

My Day—by Mike Allison

- 4:30 a.m.: Bus call (rigging call started at 2:30 a.m.). I take my bags and head to Bus 3. That is the lighting bus, but it leaves earlier, so I will take it to the venue and transfer my bags to my own bus when I get a chance after it arrives at the venue at 8:30 a.m. I try to get the cobwebs out and start thinking about the gig.
- 5:00 a.m.: We arrive at the gig. I head in with my two other audio guys. As I go in, I look at the load in. Does it have docks? Is it just dumping on the parking lot? How much space is there as I walk in? If I have been here before, I tell my guys my plan. This gig—the Bell Centre Montreal—I have been here many times. So as I walk in, I see Michell, the head sound guy, shake his hand, and say hello to the other guys I

recognize. The stage manager is in the process of dumping the stage trucks, so we have some time. My two guys head to catering. I head to the floor to have a look.

- 5:15 a.m.: I get with our tour electrician and find out where we will get our connection for power. As this tour is on the verge of needing generators, I need to check every day to make sure where we connect.
- 5:25 a.m.: I see the head rigger and ask whether he has any problems with my points. If not, then I look to see what kind of space we will be working with. We have so much gear behind the stage that space is at a premium. Today, the rear-hang PA cabs are flying right next to the back wall. We will need to get at them ASAP, or we will not be able to get them up until the lighting truss goes up, which could be a long time. I look for space to place the amps—not good today.
- 6:00 a.m.: I head to catering and grab some breakfast. It's hard some days to eat heathy; it's all greasy food. But I try.
- 6:20 a.m.: I talk with the lighting crew chief about having some of the space that is marked out for lighting repair world. (It's very large out here . . . lots of lights need work all the time.) I am glad we don't have to repair the PA as much!
- 7:00 a.m.: It looks like we may get a truck soon. I call my guys on the radio and tell them soon. Maybe . . . Take a look at space and let me know whether you want me to send it out on the floor or around the back, through the vom. [Note: Vom is short for "vomitorium." Really. An entrance into a theatre through a banked tier of seats. It comes from the Roman term for an entrance into an amphitheatre.] This place has lots of entrances. My guys think that the PA cabs should go on the floor. The amps stage right have the long push around the stage, along with the cable caddies and truss. Now the rear fill and side fill PA need to go another way.
- 7:30 a.m.: We get the PA truck and start unloading. It takes a few tries of telling guys (local stage hands) that it's not all going out on the floor. If they would listen, we could get it done without so much confusion.
- 8:00 a.m.: First truck done. Now starting on the next truck. This one is just monitors and FOH gear. But again, it's not all just dumping out on the floor. Some goes down the hall to FOH, some out where the stage is being built, some to stage left . . . We get it all in place.
- 8:30 a.m.: Truck is unloaded, and now it's time to get the sound hands and start working. We get six guys. I take two and split the other four between my guys. Dustin heads to the stage to get started on the cabling he needs to get done before everyone else gets there, and he has to fight to get into the space under the stage. Chris takes his guys and starts stage right PA. It's the harder side, with all the dimmers and cabling for the ton of lights on the floor. The sooner he gets in to get space, the better. I take my guys and start running the 2/0 feeder for power. It takes about

15 minutes. Then it's time to get on the rear fills. Stage left is a mess. All the lighting and motion cables are on the floor, right where I need to fly the rear fills. I take a look. If we can slide the cables out a little, I may be able to just get the speaker cabs in. It's a job, but we get space. It takes about 30 minutes to get them up. It's three stacks of four i3 cabs—not big. Thank god, because the space is so tight. We get them cabled and flown.

- 9:00 a.m.: The other sound guys arrive (an FOH engineer and two monitor engineers). Chris has a small problem and calls on the radio. I tell my hands to wait for me by the cable caddies and see what's up with the PA. Nothing serious—just a question about pinning for the cabs. I take a look at the cable truss and everything he has done so far just to make sure it all looks good and safe.
- Right about now I lose track of time.
- I get my hands and head out to monitor world. With the unions here, we need to use our guys to tip the two monitor boards and get them in place. After we get them in place, we head to FOH. And do the same thing... [Note what "tipping the board" means. This tour is on all big, heavy Midas analog consoles. Each weighs about 1,100 pounds in the case. "Tipping" means getting them from the ground onto their stands.]
- By now, Dustin is over at stage left getting ready to fly the PA.
- I take one guy and give him to Dustin, and the other I take to Chris.
- During this time I am listening to the room. I have been here many times, so I know how it will sound. But if it's a place that I can't remember or that's new to me, I listen to the room. When the stage guys drop something, does it echo a ton or does it taper off? As the motion guys raise the video truss, does the sound of the motors slap around where the stage will be? You get the idea—I listen to the room. Later, when I am tuning, it will help me.
- I head out to the stage left amps and wire them up with power, signal cables, and network cables. They do the same for stage right. Now I have to figure out where to run the cross-stage cable (signal and Cat-5 network cables). If I run them on the ground, they will get buried with all the lighting, video, motion, and who-knows-what cables. Not good for audio cables... So I run the cables around the first level of seats. I have 250 feet of cable, so it's not usually a problem to reach from stage left to stage right. But today it just makes it.
- I go to check on Chris. Everything is looking good. I head to stage left and tell Dustin he can go to the stage and finish what he needs to do there. I will finish flying the PA.
- With both sides of the PA floating just above the floor, I head up to the upper balcony to get a better look at the 11 motors that comprise the main PA, the downstage side-hang and rear side-hang PA, and supporting truss. From below, if one motor

stops, it is really hard to see; but from above, I can see it really easily. We have tape measures on the main, so as we go up I can get a reference. We fly the PA at 32 feet to the bottom. I get all the speakers and truss up to height. I look at level, do the pull-backs, and head to the rear speakers. We fly them independently, because they are so surrounded that this is the only way to do it.
Next to stage right. During this time we are rotating out the stagehands to take a break. Cabling the PA is always a job, with 220 feet of cables and long runs.

- About now, the lighting and video trusses are up, and the stage is getting ready to roll into place.
- Chris heads to stage right to cable up the amp racks. I head to stage left amps. I have two stagehands. We try to keep the cables falling straight down, and then we route them to the amps. Sometimes, depending on where I get the amps, we have to add the extension (between 30 feet and 60 feet). All the cables are Clair Hi-D, which is a very large-gauge cable, so it's like wrestling a snake to get it where you want it to go. Very good for low loss to the speakers, but not fun to run.
- The stage rolls into place, so it's time to put the transformer split/patch rack into place. This is where all the stage boxes (with all the mics) converge to get split to the four different places (FOH, two monitor boards, and recording mix). This is a large caddy that looks like a cable bomb when all the connections are made. It just fits under the stage with a little room to get in to make all the connections.
- As I am doing this, Chris takes the stagehands and starts running the cables for the subs that are under the stage along with the front fills. You got it right: more Hi-D. We have eight subs, 12 FF-2 (small front fill cabs), and two S-4s, the big old four-way Clair PA under the stage. So we need six of the Hi-D cables. They run under stage left and out to the amps, all 100 to 120 feet in length. More fun with the cables, as we need to run them in a very neat order and put them into large cable ramps. The Fire Marshal really likes it that way, and we need to keep anyone walking from tripping and keep gear from rolling.
- After I make the connections, I take the hands to FOH to run the real snakes. We have two runs, Bon Jovi and then the supporting act. They both are 100 meters, or about 330 feet long. They have to connect to both the transformer split and the amps.
- Finally, it's time for lunch. Most days it's a little after 1:00 p.m. It's always interesting in catering now, as we have the stagehands in eating. Some of the conversations are very interesting.
- After I finish, I head to FOH to power up the system. I turn all the FOH gear on and get all the computers running. All of the consoles are old-school analog—Midas

XL-4 at FOH and a pair of Midas Heritage 3000s at monitors. But the drive system is a very new Lab.gruppen amp Lake I/O system. I run three computers—one tablet has Smaart on it, and there are two for controlling the drive. I start the I/O software and wait as it talks with all the amps. (Each amp has a Lake digital controller in it, which is on anytime the amp has power to it, even if the amp is off.) It takes about five minutes (on a good day) for all the modules to check in and see whether they are in the same state as the controller thinks they should be. If they are not communicating, or if they are showing offline, then I have to figure out what's up and how to fix it.

- I get them all talking: 60 amps, 208 amps channels. It's time to turn them on. With the new amps, the real way to do this is with the software. I go to the PLM page and hit All On. Wait 10 seconds, and they all start turning on. The software reports all is good. Now it is time to check the system. Again, new amps—the only way to really do this is with the software. So I start to mute the different outputs that go to the Low, Mid, and Hi, Sun S-4, and FF-2. With all the channels, this takes a few minutes. I turn on the pink noise and start unmuting the outputs one at a time to make sure every component is working and all connections are working. With a large system, you can see this takes a while.
- All systems go. Dave (Eisenhower—FOH engineer for Bon Jovi) comes out, and we start tuning the system. We have a good system, and we know what we want, so it's a relatively easy time with the initial tune. Dave and I have the same feeling about the PA. If you have good sounds coming in, and you know the PA sounds good, then the tune should be easy. We do minor tweaks at FOH. Get the main PA delayed to the subs. Then I fire up the second tablet and start the VNC program, which is basically a server on the main tablet and a client on the second tablet. We run a remote desk on the wireless to walk the room. It runs Wi-Fi at 2.4 GHz and 5 GHz. As we are running a 360-degree system, we have to check from top to bottom, front to rear. So we drive all the other guys crazy with the same song repeating for a while. The goal is to reproduce the sound at FOH all over. Most days I believe we do a good job. The system is so separated out that it makes this easy.
- System tuned. It's time for line check. I relax for a few and start planning the load out. It has been in the back of my mind as we load in. When I am placing anything, flying the PA and loading in, I am thinking about how the load out will go. If I do that or place that there, how will that affect the load out?
- Bon Jovi line check done. It's time for support check. This is always one of the hardest times for me. Depending on the act, most have not ever had any time in a big arena, so they are a little out of their experience level. How do I keep them happy, keep the gear safe, and keep the main act and production happy? Some of the mixers think that LOUD is good, but a big arena with loud in it never works. The politics for this are way too long to go into here.

- We get finished and wait for Bon Jovi sound check. Most days it's a few songs. We work on anything new they are doing tonight.
- 5:00 p.m.: Bon Jovi sound check on stage.
- 6:30 p.m.: Doors. Start the walk-in music. Time for dinner.
- 7:15 p.m.: Run the house safety announcement.
- 7:30 p.m.: Support act on. Keep their engineer "in the program." Help if I can; otherwise, just count the time.
- 8:15 p.m.: Set change. Quick line check. Get ready for the show.
- CD in the recorder.
- Radio announces four minutes to band. Start the intro.
- The show starts. I wait for one song to let the band settle and Dave to get into the groove. Then I take the secondary tablet and walk the arena. Usually it's just small changes—gain up here or down there.
- End of show. Load out. The best way to describe it is controlled chaos.

Note a few things: His day started at 4 a.m., and lunch was at about 1 p.m. That's a nine-hour day, and he is not even half finished yet. On the show I saw, the band finished around 11 p.m. A typical load out is two hours or so. This is a really big tour, so it is likely closer to three . It is a good 22-hour day.

And there's another show tomorrow....

27 Just Because You Can Doesn't Mean You Should

The title of this chapter may be the best advice anyone can give you, and it doesn't matter whether the advice is technical or just about the way you comport yourself on the gig. For example:

- The audio systems in use today are very powerful, and it is possible to run volume levels exceeding 110 dB. Don't.
- The digital systems that are increasingly the norm allow an engineer to put processing on every channel. Compression, reverb, and any other effect you can think of are as close as the nearest computer plug-in. And you can stack five or six inserted processing units on any channel. Don't.
- Given the right combination of stimulants, you may be able to party like a rock star and still be able to get through the next day's gig—for a while. Don't.

This next part is something that has been around for years but never before published. Titled "The Road Bible," it was written by a touring sound guy when—after a 12-hour-plus pre-gig call—he was unwinding on the bus and a guest came aboard, looked at him, and said, "I want your job . . . ."

So You Want My Job?
By Roadie "M"

Don't anyone take this offensively. I'm taking time from figuring out my float to give some insight to anyone who wants to work on the road. Read on . . .

Do you have a family, a pet, a wife, kids, a home, a disgusting habit, a desire for lots of sleep, a diet that consists of water and tofu, motion sickness, the inability to lift more than 75 pounds (just to name a few)?

If you answered yes to any of these, stay at home, because you won't make it out here, and what you had at home will probably be gone when you get back.

While you're thinking of your mate getting "entertained" by your neighbor or Fido running into traffic or your kids telling you about your "long-lost brothers," who sleep over while you're gone, all after whining to us about your $1,000 cell phone bill, I am worrying about you dropping a shackle from the grid, tipping over a PA stack, electrocuting a lighting guy, or puking in someone's bunk, which may lead to you getting yourself in deep trouble.

There's nothing worse than having to pick up a slacker's slack (especially when it's preventable).

Think you're a good candidate? Hardcore? Ready for battle? Read on . . .

Schools
Schools are a good foundation, but no school can simulate an 18,000-seat building or teach you how to lift a console, pull cable, stack, power up, and repair gear without hurting yourself or someone else, all while making it sound good and look good all while getting it there in one piece. If you have the money, go for it. But don't think for a minute that a sound or lighting company is going to make you a trade-mag-worthy engineer overnight. After you spend 30 grand and get your first gig, guess what? You'll still be stacking PA and trusses, pulling cable, loading trucks, and driving trucks, buses, and so on. There is no substitute for experience, but in a school you will learn some basics. Knowing what to do when something goes wrong will come with experience—unless you're ignorant or lazy.

The Big Green Book of Rock and Roll
Page 1: The gig comes first.

Page 2: You will never see anything in this book that tells you this job is fair.

Page 3: Don't be late. If you are, bring a copy of your obituary.

Page 4: Treat people well on the way up. If you don't, those you stepped on will step on you as you come back down.

Page 5: The band will always win.

Page 6: Know what to do when something goes wrong. Or else.

Page 7: Be nice. Being a jerk will only get you a window or aisle seat—one way.

Page 8: No one cares how you did it with XYZ band. There's a right way, a wrong way, and our way.

Page 9: Don't fix it 'til it's broke.

Page 10: The wheel has already been invented; sometimes it needs new tires or a little bit of air. Have a workbox and an air hose ready and know how to use them. Trying to reinvent the wheel may cause you to get locked in a workbox.

Page 11: The rumor you started will end. See Page 7.

Page 12: Everyone gets paid to do his job. Want more money? Get another job or be the best at what you do. Learn your boss's job.

Page 13: Trust and respect are earned.

Page 14: Thou shalt not lie. Thou shalt shut up. Eventually the truth will come out. Earn respect or get a choice of seats (see Page 7—again.).

Page 15: Don't be stupid. Check and double-check, or someone may not live to see their next paycheck.

Page 16: Tour managers/production managers do exactly what their title says. They arrange travel as well. And always have cash and credit cards. Anger them, and you'll find out what else they can do.

Page 17: Nothing is secret or sacred on the road. We will find out eventually.

Page 18: Ask if you don't know. Just don't ask the wrong person.

Page 19: What happens on the road stays on the road.

Page 20: Understand politics but don't get involved with them.

Get the picture? Eventually you will . . . maybe. Keep reading . . .

If you really want to work in the touring industry, find a touring-related job.

Go to a sound company, a lighting company, a trucking company, a pyro company, or a bus company and apply for a job. Go to a local show and get on the "call," get hired, and pay your dues, and you will be partially on the road to success. Keep quiet. Listen and learn and have no fear of manual labor. No one will hire you from job corps, the newspaper, or the lottery. If you want this job, *you* have to get it. It rarely gets handed down.

The stuff you put up with in the beginning will educate you for later—school or no school. We don't want snivelers, whiners, or crybabies. Ex-military folks do well on the road and adapt quickly. Want to know why? Discipline, familiarity with lousy conditions, knowledge of teamwork and organization, knowing how to go through a chain of command, and above all, learning to be an expert in their field and understanding unit camaraderie.

There is no grievance board, company nurse, sick days, or foo-foo baskets on birthdays. The tour is the nation, with different ranks, rules, and policies. Many are unwritten. This isn't just a job; it's a different way of life. Be good at something. If that something is desirable, you will always be working. Don't sell yourself short. If you can't live on what someone wants to pay you, find another job.

You want my job? Get up and do something. Don't just sit there; do something. Open a phonebook, look in a newspaper, in *Pollstar,* or at a tour schedule online and *do something!*

On another note, the rumors are true. "Touring can be like boy scout camp with girls, liquor, and [fill in the blank]."

After the work is done.

Now go!

When I was asked to do this book, I put a question to the members of ProAudioSpace.com. I asked them what they wished the "new guy" who just came out of school knew before he got on the gig. I got dozens of responses, and not one of them had anything to do with gear or technology.

One of the most common requests was that new folks learn the right way to coil or wrap cable. It's called "over-under," and it is really hard to illustrate in a book. Just remember that you never wrap cable in a loop between your hand and elbow. More quickly than anything else, it will tag you as knowing nothing. If you don't know how to "over-under" a cable, ask. And then practice at home.

Another common issue was the inability of many newbies to troubleshoot. We have become a "throw it away" society. Most of us in the audio biz grew up when there were still things like "shop" classes in high school, and we learned the basics of an electrical circuit and how to build and repair one. Today it often costs more to repair an electronic product than it does to just replace it. The result is a serious lack of ability in the troubleshooting department.

First, many problems can be avoided with proper preparation. For example, check all snakes with a condenser mic and make sure there is signal on every channel before you even leave the shop. Do it again before pinning the stage, and you will avoid problems that will take longer to find later in the process.

Here is how I troubleshoot. It comes down to a process of elimination to find the problem. You have to find it before you can fix it. Say there is a mic onstage that is not sending signal to the board. In a simple setup, it could be the mic, the cable, the snake channel, or the console channel. Or it could be something as simple as the phantom power not being engaged on a channel using a condenser mic. Here are the steps I take.

1. Check the board to make sure the head amp is up, the channel is not muted, another channel is not soloed, and the phantom power is on if it is a condenser mic. Also check where the snake outs connect to the board. A connector may not be inserted all the way.
2. Replace the mic with a simple dynamic that you know is working. Make sure to mute the channel or bring the gain all the way down before taking this step.
3. *Still using the mic you know is working,* replace the cable.
4. If it is still not working, then it is either the snake or the console. (Remember, this is a simple setup.) The easiest thing to do is to plug that same working mic directly into the console. If it still does not work, replace the cable, just to be sure. If it is still not working, figure that the console channel is bad. Tape it off and worry about it after the gig. Repatch that input to another channel.

5. If the mic works plugged directly into the channel, then it is a problem with the snake. Repatch it into another channel.

The same basic process should work no matter what the problem. If one side of the PA is humming, switch the outputs (if that is possible). If the other side starts to hum, then the problem is at the console. If the same side hums, then it is somewhere between the console and the speakers. Just keep moving down the line and isolating individual links in the chain until you find the one that is bad. Always have spares. If the faulty piece goes on the gig, the chances of having the time to fix it are not good. That gets done at home or at the shop after the gig is done. Refer to the signal flow chart at the end of Chapter 19 if you are confused.

Next request from the tribe: Know the lingo. Do you know what a spanner, a torch, and a stinger are? In order, that would be a wrench, a flashlight, and an electrical extension cord. But no need to reinvent the wheel here; there is a great Lingo section on Roadie.net. Learn it.

The following sections provide some more of the assembled wisdom of people who have been doing this for a long time.

Dave Shirley

My first day in a Sound Reinforcement 101 class was in September, 1981, at Belmont College (now University) in Nashville. The first day of audio class, they taught us that the most important rule to remember is, "No matter what happens, it's always the soundman's fault!" If you can't remember anything else, remember that rule, and it will make those "learning experiences" at first gigs make more sense. If the singer loses his voice, it's the soundman's fault. If the PA died, it's the soundman's fault. If the audience couldn't hear the lyrics in the cheap seats, it's the soundman's fault. If they hated the band, it's the soundman's fault. I'm sure you follow the pattern. It's somewhat funny, yes, but also completely true. All kinds of stuff will get blamed on you, deserved or not. Deal with it, learn from it, and move on.

The second thing they taught us the first day of class was the six questions every sound guy sound should ask about a gig—and this works in almost any business interaction, especially if you are self-employed, as most of us are. The five Ws and an H: Who? What? When? Where? Why? And how much? If you can't answer those six questions, you didn't get all the info you needed to book that gig! Get all the pertinent details in the first contact with the client. It will make you appear professional and cut down on playing phone tag and sending lots of emails back and forth. Not that there aren't follow-ups with additional info and variables, but if you know the basics, you can add it to your calendar with confidence.

Oh, the extra-credit question on the end of semester final: "Whose fault is it?"

Jordan Wolf

1. Learn how to effectively communicate with people.
2. Learn how people perceive your communication toward them.
3. Learn the various ways to wrap cables. (Different companies have different preferences.)
4. Be willing to work long hours for little pay, respect, or acknowledgment.
5. You're *never* too important for sweeping floors or cleaning fader caps.

Ian Silvia

Someone should kindly explain to them that if they work for a sound company, they might not get to mix bands very often. It seems like most bands (including openers) have their own "engineers" these days. I think it might have something to do with the serious number of kids these schools churn out. They get out of sound college and go on tour with their buddy's band for $50 a day and a ham sandwich.

If you like knobs, faders, nice clothes, and catering, find a gig working for a band. If you like feeders, motors, gig butt, and gas-station egg salad sandwiches, get a gig with a sound company.

I prefer the latter.

Mike Reeves

I notice a lot of new guys have a habit of patching the cable from the microphone first, making a whole bundle of cable by the snake. Obviously, we know to start at the snake and neatly drop the cable underneath the microphone stand.

Considering when techs get out of school, chances are they aren't going to be behind a console anytime soon. Since they are learning about how to run the consoles, they need to know what they are getting into once they arrive. So load-in/load-out etiquette... Learn how to properly lift gear. Learn spatial relations on how to load trucks correctly. Suggesting they get a chauffeur's license might be good, because the new guys are usually the ones who have to make the deliveries in the box trucks. Learn that you work as a team, and *never* leave your man hanging. Nothing angers a tech more than the new guy vanishing during a load in/load out.

D.V. Hakes II

In my neck of the woods—Hawaii—there are many decent A-1s. Basically, someone has to die or retire before these kids even get a shot at touching the board. We get an onslaught of them every now and then, but none of them stays for long (which is a shame in some cases), because they never get the chance to mix, and they don't know how to do anything else (or they don't want to). They want to bypass the five to 10 years of being an A-2 and go directly to being an A-1.

They need to teach these guys/gals (or they need to learn as quickly as possible) mic placement; speaker placement; speaker rigging; mic type for different applications; how to take amps apart and put them back together again; how to mark the cable ends and the snake heads (for everything); how to set up, run, and fix different COMM systems and how to get them to talk to each other; and so on.

Having worked with 30 or so of these kids over the last 15+ years, I've learned to let them know straight up that they aren't going to touch a board until they know how to set up and fix everything else (as well as show their work ethic, and so on). I know it's harsh, but that's the way it works around here. We have more experienced A-1s than we have work. We also have a list of A-2s who have put in the time and have the knowledge, who are waiting (and very ready) to step up.

Jeanne Knotts

1. Calm down and pay attention.
2. Know your frequencies. If you didn't learn this in school, then you can teach yourself with an audio training CD or an EQ in line with your stereo.
3. Don't lie your way into jobs. It generally backfires, and you stop getting booked.
4. Be good to your cables and tape them down.
5. Learn to solder and learn basic repair skills, because you will usually need them in a high-stress situation, on the fly, and with no light. Keep your soldering iron on during the show, just in case.
6. Have backups for all your tools *and keep them locked up!* Unless I specifically need something, I generally only carry what I can fit in my pockets and on my belt.
7. If you think you will fail, you're right. See #1, keep a positive attitude, and stay busy or ask what you can do. Sure, there's a lot of hurry up and wait, but if people think you're lazy or an idiot, they'll book someone else. There is a difference between not knowing something and being an idiot. Learn it.
8. Smoking and swearing are hot-button issues now, and unless you're good friends with the talent, you're better off keeping the relationship professional. Also, you never know where or how you will run into someone, so try not to step on people if you can help it.

Steve McCarthy

It is easy to spot the newbies when they try to wrap cable. They always wrap it around their palm and elbow. They lift with their back bent over instead of with their legs.

"Keep your hands out of your pockets, kid."

"Don't stand there gawking with your jaw hanging down."

"Ask questions. No one is going to preach audio to you."

"Put your gloves on and stop crying about that cut on your hand."

"You want to patch the snake head? Try counting from 1 to 10 and 10 to 1 at the same time, for starters."

"Stay in your department. Don't let me catch you drifting off with your friends in props."

"You want to reinvent the wheel, kid? Save it for the day you become a department head."

Final Words

A final thought: We have so much control at our fingertips that it is easy to forget that sometimes too much choice is a bad thing. It is not a coincidence that great live records were released on a seemingly monthly basis in the '70s and '80s, and there are very few of them anymore. But most shows have the ability to record and even do full multitrack sessions on every gig.

Those great live records were made on gear that baby bands won't put on a rider. The Beatles played Shea Stadium with gear that no self-respecting bar band would use today.

The thing is that music was the primary concern when those records were made—not technology. Whatever you do, remember that you are there in the service of the music or whatever the performance is. The most successful gig is the one where no one knows you are there. It's not about you. It's not about the gear. It's about the performance.

And never forget that extra credit question: Whose fault is it?

Now get up. Doors are in 30, and there's still work to be done.

Index

C

D

I

J

K

L

M

U

V

W

X

Y